EDI DEVELOPMENT POLICY CASE SERIES
ANALYTICAL CASE STUDIES • No. 4

Morocco
Analysis and Reform of Economic Policy

Brendan Horton

The World Bank
Washington, D.C.

Copyright © 1990
The International Bank for Reconstruction and Development / THE WORLD BANK
1818 H Street, N.W.
Washington, D.C. 20433, U.S.A.

All rights reserved
Manufactured in the United States of America
First printing June 1990.

The Economic Development Institute (EDI) was established by the World Bank in 1955 to train officials concerned with development planning, policymaking, investment analysis, and project implementation in member developing countries. At present the substance of the EDI's work emphasizes macroeconomic and sectoral economic policy analysis. Through a variety of courses, seminars, and workshops, most of which are given overseas in cooperation with local institutions, the EDI seeks to sharpen analytical skills used in policy analysis and to broaden understanding of the experience of individual countries with economic development. In addition to furthering the EDI's pedagogical objectives, Policy Seminars provide forums for policymakers, academics, and Bank staff to exchange views on current development issues, proposals, and practices. Although the EDI's publications are designed to support its training activities, many are of interest to a much broader audience. EDI materials, including any findings, interpretations, and conclusions, are entirely those of the authors and should not be attributed in any manner to the World Bank, to its affiliated organizations, or to members of its Board of Executive Directors or the countries they represent. The World Bank does not guarantee the accuracy of the data included in this publication and accepts no responsibility whatsoever for any consequences of their use.

Because of the informality of this series and to make the publication available with the least possible delay, the typescript has not been prepared and edited as fully as would be the case with a more formal document, and the World Bank accepts no responsibility for errors.

The material in this publication is copyrighted. Requests for permission to reproduce portions of it should be sent to Director, Publications Department, at the address shown in the copyright notice above. The World Bank encourages dissemination of its work and will normally give permission promptly and, when the reproduction is for noncommercial purposes, without asking a fee. Permission to photocopy portions for classroom use is not required, though notification of such use having been made will be appreciated.

The backlist of publications by the World Bank is shown in the annual *Index of Publications*, which is available from Publications Sales Unit, The World Bank, 1818 H Street, N.W., Washington, D.C. 20433, U.S.A., or from Publications, Banque mondiale, 66, avenue d'Iéna, 75116 Paris, France.

At the time of writing, Brendan Horton was a research associate at the African Studies Center, Boston University.

Library of Congress Cataloging-in Publication Data

Horton, Brendan, 1944-
    Morocco : analysis and reform of economic policy / Brendan Horton.
        p.   cm. -- (EDI development policy case series. Analytical case studies ; no. 4)
    Includes bibliographical references.
    ISBN 0-8213-1432-7
    1. Morocco--Economic conditions. 2. Morocco--Economic policy.
I. Title. II. Series.
HC810.H67   1990
338.964--dc20   89-77569
                                                                CIP

EDI Catalog No. 400/039   ISSN 1013-333X

Contents

Preface *vii*

1. Introduction *1*

2. The Moroccan Economy, 1970-87 *3*

   Economic Structure *3*
   Principal Economic Developments, 1970-87 *5*
   Summary of Stabilization and Adjustment Since 1975 *12*
   Government Revenues and Expenditures Including
     Debt Amortization *13*
   Conclusion *20*

3. Methodology *22*

   A Framework for Comparing Approaches to
     Adjustment and Stabilization *22*
   Measurement of Production Incentives *27*
   Tax Policy *29*
   Interest Rate Policy *29*

4. Economic Policy Reform, 1973-82 *31*

   Expansion and Stabilization, 1973-79 *31*
   Stabilization and Adjustment, 1980-82 *34*
   A Comparison of Approaches to Adjustment, 1973-82 *39*

5. Adjustment, 1983-87 *41*

   IMF Operations *41*
   World Bank Operations *47*
   Conclusion *88*

6. Lessons from the Case Study *91*

    The IMF's Approach to Stabilization and Adjustment *91*
    The World Bank's Approach to Stabilization and Adjustment *95*
    Conclusion *99*

References *102*

Annexes *105*

    1. Synopsis of Balance of Payments Support, 1970-87 *107*
    2. Tables *111*

Tables in Text

    2.1 Composition of GDP by Sector, Selected Years, 1969-87 *4*
    5.1 Effective Protection Coefficients by Market *49*
    5.2 Domestic Market EPCs for Selected Industries *50*
    5.3 Net Effective Protection Coefficients by Market *52*
    5.4 Structure of Import Duties and Receipts *56*
    5.5 Issues, Objectives, and Corresponding Policy Measures in the ITPA1 Program *59*
    5.6 Proposed and Implemented Rates of the Special Import Tax and Stamp Duty, 1983-88 *75*
    5.7 Import to GDP Ratio by Import Regime, 1983-87 *80*
    A.1 Main IMF Facilities *107*
    A.2 IMF Operations in Morocco, 1970-86 *108*
    A.3 IMF Operations in Morocco by Type, 1970-86 *109*
    A.4 World Bank Operations in Morocco *110*

Figures

    2.1a Current Account and Budget Deficit, 1970-87 *6*
    2.1b Current Account and Resource Gap, 1970-87 *7*
    2.1c Current Account, Domestic Savings, and Investment, 1970-87 *8*
    2.1d External Current Account, 1970-87 *8*
    2.2a Composition of Investment, 1970-87 *9*
    2.2b Nongovernment Investment, 1972-87 *9*
    2.3 GDP and External Debt, 1972-87 *10*
    2.4 Monetary Authority Reserves, 1970-87 *10*
    2.5 Net Foreign Assets of Banking System, 1970-87 *11*
    2.6 Terms of Trade and Exchange Rates, 1967-82 *11*
    2.7 Nominal and Real Exchange Rates, 1977-87 *12*
    2.8 Government Receipts and Expenditures, 1971-87 *14*
    2.9 Government Expenditure by Category, 1971-87 *14*
    2.10 Noninterest Recurrent Expenses, 1971-87 *15*
    5.1 Factor Service Balance, 1970-87 *77*

5.2 Exports of Goods and Nonfactor Services, 1970-87   78
5.3 Exports of Goods, Including Subcontracting, 1970-87   78
5.4 Imports of Goods and Nonfactor Services, 1970-87   79
5.5 Merchandise Imports by Category, 1970-87   80
5.6 Global Composition of Taxes, 1972-87   84
5.7 Indirect Taxes and Consumption Subsidies, 1972-87   87

Preface

This case study reflects a decade of work on economic policy analysis and reform in Morocco. My work began in 1979 with a four-year residence in Rabat, where I served at the Ministry of Commerce and Industry as principal investigator for a joint research project by the World Bank and the Government of Morocco into the structure of incentives in Morocco's industrial sector. I later participated in the preparation, negotiation, and supervision of the resulting Bank-lending operations that have supported the Moroccan government's adjustment programs. This opportunity led me to ponder the relative merits of different approaches to stabilization and adjustment, particularly those of the International Monetary Fund and the World Bank.

I have had three special concerns: how to design and implement structural reforms at the microeconomic level that are consistent with the exigencies of macroeconomic stabilization; how to understand the political economy of the reform process; and how to explain to nontechnical policymakers the technical economic concepts that underlie growth-oriented structural adjustment programs that must take account of tight fiscal constraints. This case study is the result. As such, it reflects the contributions of many of my colleagues during the past decade. I thank them all, but errors of fact, analysis, and interpretation are my sole responsibility.

At the World Bank, Bela Balassa, Michael Carter, François Ettori, and John Shilling have given me valuable guidance and encouragement over the past decade. My former colleagues in Rabat at the ministries of Commerce and Industry, Finance, and Economic Affairs significantly influenced my thinking about the politics and economics of structural adjustment. In particular, I would like to thank Azzedine Guessous, Minister of Commerce and Industry; Taib BenCheikh, Minister of Economic Affairs; Hassan Abouyoub, Director of External Trade; Abderazzak El Moussadak, Director of Industry; and Jai Hokimi, Director General of Customs.

The detailed comments of P. HariPrasad, Martha Starr, and Barbara Lewis as well as those of two anonymous reviewers on early drafts were greatly appreciated. I also gratefully acknowledge the assistance and exceptional patience of David Davies, project director within the Economic Development Institute of the World Bank; Barbara de Boinville, my editor; and Sonia Hoehlein, who typed the manuscript. To my wife, Darnella, I am indebted for her unflagging support and encouragement during this long endeavour.

Brendan Horton

1
Introduction

This is a case study of policy analysis and reform regarding growth stabilization and adjustment in Morocco from 1973 to 1987. Since 1978, successive governments have attempted to address Morocco's heavy external debt through a series of stabilization programs supported by the International Monetary Fund (IMF) and sectoral adjustment programs financed by the World Bank. Together they have provided between US$1.5 and US$2 billion of financial assistance (see Annex 1, Tables 2 to 4). In addition, Morocco has benefited from substantial rescheduling by its official and commercial creditors.

The debt arose from large national defense expenditures as well as from projects and programs to raise the growth rate and attack the social problems of underdevelopment and absolute poverty. Stabilization and adjustment has had considerable domestic costs. Cuts in private and public capital formation impeded the realization of the government's ambitious growth and development objectives.

Politically, this period of attempted retrenchment has not been easy. Consumer purchasing power has decreased sharply, which on several occasions prompted social unrest and even food riots. During the course of adjustment, the government changed its policy toward foreign investment. Once ambivalent toward foreign investment, it now shows a new openness.

The Moroccan economy combines a vibrant modern industrial sector and a more traditional agricultural sector. From a judicial standpoint, administrative law reflects the country's historical partition between France and Spain during the Protectorate as well as the wider multinational tutelage implicit in the Act of Algeceras.

Policy analysis and reform in such an environment are complex issues. While stabilization suggests unchanging institutions, adjustment does not. Adjustment necessarily addresses wider issues—economic and noneconomic alike—that cover macroeconomic and microeconomic questions as well as regulatory and administrative reform. Adjustment programs of this nature are ambitious and require both economic and political analysis. The success of institutional reforms is contingent upon the creation of new alliances among a multitude of interest groups.

The Moroccan case is especially interesting in this regard because the World Bank's sectoral adjustment programs have been built on this premise, beginning with a three-year joint research project on the structure of price and nonprice incentives in the industrial sector. Subsequently, this project was replicated in agriculture. At the same time, however, stabilization—in particular, budget deficit reduction under the auspices of IMF programs—has been equally indispensable. The same is true of debt relief.

This case study examines Morocco's attempts at stabilization and adjustment over the past decade, with a special focus on IMF and World Bank operations. Its primary objectives are as follows:

- to describe each institution's operations, capturing the essence of their analysis as well as government reactions to it;
- to better understand how the two institutions' economic approaches differ and how they may be better combined in the future;
- to document how consensus was reached on programs that were finally adopted; and
- to grasp the main elements of the internal policy reform process so that subsequent stabilization and adjustment programs may take better account thereof.

Achieving these objectives requires a combination of analytical economic history and comparative economic analysis of IMF and World Bank approaches. An appreciation of the various Moroccan actors' reactions to these approaches is also needed.

The case study is organized as follows. Chapter 2 presents a historical overview of economic developments from 1970 to 1987 and stabilization as well as adjustment attempts since 1975. It concludes that 1983 is a natural dividing point for the subsequent analysis. Chapter 3 presents a synopsis of the various methodological approaches that are used in Chapters 4 and 5, which cover 1973-82 and 1983-87, respectively. Chapter 6 succinctly summarizes the preceding four chapters and draws lessons therefrom. Although it develops ideas presented earlier in the book, it is intended to be relatively self contained.

2

The Moroccan Economy, 1970-87

The purpose of this chapter is to provide an overview of principal economic developments in Morocco between 1970 and 1987, focusing in particular on the external current account, the budget deficit, and the exchange rate. A synopsis of stabilization and adjustment attempts since 1975 follows, where government receipts and expenditure, including debt amortization, are discussed in detail. In this way, insights will be gained into government policy as well as the importance of debt rescheduling provided by foreign creditors.

Economic Structure

Morocco is a lower middle-income country with a per capita income of about US$600 in 1985. Its economy is well diversified with a large and varied natural resource base. The most important natural resource is rock phosphate. Morocco has about 75 percent of the world's known reserves. An increasing share of production is processed domestically into phosphoric acid and fertilizers for export. The nonchemical industrial sector produces a wide range of goods, including leather and mechanical and electrical products. The agricultural sector, which employs 44 percent of the labor force, is predominantly rainfed and oriented toward food production (cereals, edible oil seeds, potatoes, and pulses) and livestock (sheep, poultry, and beef). Irrigated agriculture accounts for about 10 percent of total arable land but contributes over half of the sector's value added (mainly citrus, vegetables, and sugar) as well as two-thirds of its exports. The fishing sector (canned sardines, fresh and frozen white fish) is of growing importance.

In general, Morocco's exports reflect its diversified economy. Phosphates, textiles, electronic goods, handicrafts, citrus fruits, vegetables, and canned and frozen fish are exported. Tourism is also important in the external sector as are remittances from the many Moroccans working and residing abroad.

Table 2.1 shows the composition of GDP by sector for the 1969-87 period. Drought is the main factor underlying the fluctuations of agriculture's share. It was especially important in 1977 and again in 1981, when agricultural production declined by 23 percent compared with 1980. Inadequate rainfall continued to be a significant problem until 1985-86, when bumper harvests occurred for the first time in over a decade. Industry's share, 27.7 percent in 1969, rose to almost 35 percent in 1975 with the quintupling of world phosphate prices. Note also the rapid growth of the construction subsector between 1969 and 1975. Growth continued into 1977, when construction's GDP share reached an all time high of 9.4 percent. The main cause was investment in infrastructure (roads, dams, etc.). By 1983, however, industry's GDP share had declined to 31.8 percent; in the next two years there was a 2 percent rebound. The service sector has fluctuated between 41.1 and 47.8 percent of GDP.

Table 2.1 Composition of GDP by Sector, Selected Years, 1969-87

Sector	1969	1974	1975	1980	1983	1985	1987
Agriculture	19.3	20.5	17.3	14.9	17.0	18.4	18.6
Industry	27.7	34.9	34.7	33.6	31.8	33.8	31.0
Energy	3.5	2.9	2.6	3.6	3.4	3.9	3.6
Mining	3.6	12.8	9.1	5.5	4.3	4.7	3.4
Manufacturing	16.7	15.6	16.6	17.5	16.9	16.9	18.5
Construction	4.0	3.7	6.5	7.0	7.1	6.6	5.6
Services	47.8	41.1	43.8	44.6	44.4	43.2	44.4
Transportation	19.2	17.0	18.1	19.1	19.2	19.5	20.1
Commerce	18.3	15.1	15.4	12.5	12.0	11.6	12.0
Government	10.3	9.1	10.2	13.0	13.2	12.1	12.3
Import Taxes	5.2	3.5	4.2	7.0	6.7	4.6	6.0
GDP	100.0	100.0	100.0	100.0	100.0	100.0	100.0

GDP growth has been irregular and on occasion less than population growth, itself close to 3 percent per year. Growth was greatest between 1973 and 1977, which may be attributed to the 1974-76 boom in phosphate prices and the associated expansion of the mining, construction, and energy sectors. A second major influence was the increase in investment, particularly public investment, to 34 percent of GDP in 1977. Growth declined intermittently between 1978 and 1980 in line with the stabilization efforts made necessary by the 1974-77 boom. In 1981, the growth rate turned negative in the wake of a 23 percent drought-induced decline in agricultural production relative to the preceding year. In turn, the improved performance of 1982 was only partly explained by agriculture's rebound; other factors included an upturn in construction, energy, and services. This overall improvement was short-lived. A major stabilization program begun in 1983 retarded growth until 1985. Since then there has been a moderate upturn because of above average harvests as well as increases in output of energy, construction, and industrial exports.

Analysis of growth rates by sector provides some useful insights into the economy's overall growth. First, both the mining and agricultural sectors exhibited a tendency to large interannual variations in growth performance. In mining this was a major factor only in 1976-77, but in agriculture there was negative growth in eight out of the fifteen years analyzed. Second, the service sector in general and the government subsector in particular recorded the highest average as well as the least variance in growth rates.

Principal Economic Developments, 1970-87

At the beginning of the 1970s, Morocco, like many other countries, faced immense challenges and difficulties related to economic and social development. Many of these were addressed in the 1973-77 Economic Plan. At the same time, a Moroccanization drive was undertaken in 1973, especially in the agricultural sector and to a lesser extent in the industrial sector, through the promulgation of investment and export codes. These tended to restrain foreign investment in Morocco on the one hand and to encourage import substitution via nascent Moroccan industrialists on the other.[1] Export growth was expected to result primarily from domestic processing of the country's principal resource, rock phosphate, as well as from traditional agricultural exports.

In the wake of the oil price shock of October 1973, world prices of phosphates quintupled. This led to a positive net terms of trade shock for Morocco in the sense that the additional foreign exchange earnings from phosphates more than compensated for the additional outlays on imports (namely, crude oil, wheat, sugar, and other foodstuffs). Indeed, the gain was so large that the 1973-77 Economic Plan's objectives were revised upward. A major increase was planned in public capital expenditure. At the same time, the government decided to subsidize sugar, petroleum, and other commodities whose international prices had substantially increased. Finally, in 1975, military expenditures escalated as a result of the repossession of the former Spanish Sahara.

On account of these factors, domestic expenditures on consumption and investment rose sharply from 1974 to 1975 even as the boom abated in phosphate prices. By 1976, the price of phosphate had descended to its 1973 value in real terms. Nevertheless, the government did not cut back the investment program. Consequently, large and potentially enduring external current account and budget deficits emerged in 1976 and 1977.

Specifically, the external current account changed from a surplus of 3.1 percent of GDP in 1974 to deficits of 14.6 percent of GDP in 1976 and 16.5 percent in 1977. Similarly, the overall budget deficit increased from 3.9 to 18.1 percent of GDP between 1974 and 1976, before improving marginally to 15.8 percent in the following year. In large part, these deficits were financed by recourse to heavy foreign borrowing that increased from 12.5 to 37.9 percent of GDP in the three-year period from 1974 to 1977.

By 1978, it was apparent that strong stabilization measures were needed to forestall a serious external debt crisis: within the preceding decade, debt service had soared from 5.6 percent of GDP in 1975 to 18.8 percent. Stabilization measures somewhat improved the external current account and overall budget deficits in 1978 and 1979.

By the time of the second oil shock in October 1979, Morocco had not yet adjusted to the first. As in 1974, there was an upward movement in the price of phosphates but of much smaller proportions and not enough to offset the growth of imports from

---

1. The Moroccanization programs resulted from the political desire to repossess the lands and assets acquired by the French during the Protectorate period from 1912 to 1956. Expatriate farms averaged 170 hectares and Moroccan-owned farms 8 hectares. Moreover, expatriate farmers enjoyed preferential access to the most productive rainfed and irrigated lands, credit, and profitable metropolitan markets. Moroccan farmers were mainly limited to the subsistence crops sold primarily on the domestic market. Also their tax burden was heavier.

1979 to 1981. This import growth was attributable to a combination of factors—higher prices for crude oil, more capital goods imports associated with the implementation of the 1981-85 Economic Plan, and large imports of cereals in response to the decline in agricultural production resulting from the drought of 1980. As a result of these factors, the current account deficit worsened to 12.5 percent of GDP in 1981. The Treasury deficits that year represented a sizable 14.5 percent of GDP.

Expansionary policy was continued in 1982, although there was not a significant deterioration of the resource gaps. External debt at the end of 1982 had risen to almost 68 percent of GDP, more than 50 percent above its end-1979 level. Debt service amounted to 37.1 percent of exports of goods and nonfactor services. Nevertheless, a further expansion was planned for 1983. It would again be fueled mainly by heavy foreign borrowing.

A foreign exchange crisis resulted in March 1983. Official reserves were not enough to service the external debt and provide the real economy with the merchandise imports it needed. Emergency financial assistance was solicited from external creditors (bilateral, multilateral, and commercial) in the form of new loans and the rescheduling of existing debts.

Since 1983, the external current account and overall budget deficits (after rescheduling) have improved substantially on account of the government's adjustment efforts, good agricultural harvests, debt rescheduling, and the 1986 decline in the world price of petroleum. By 1987, the external current account showed a surplus of 1 percent of GDP, while the budget deficit had declined to about 6 percent of GDP. The ratio of debt to GDP had decreased to 120 percent of GDP after

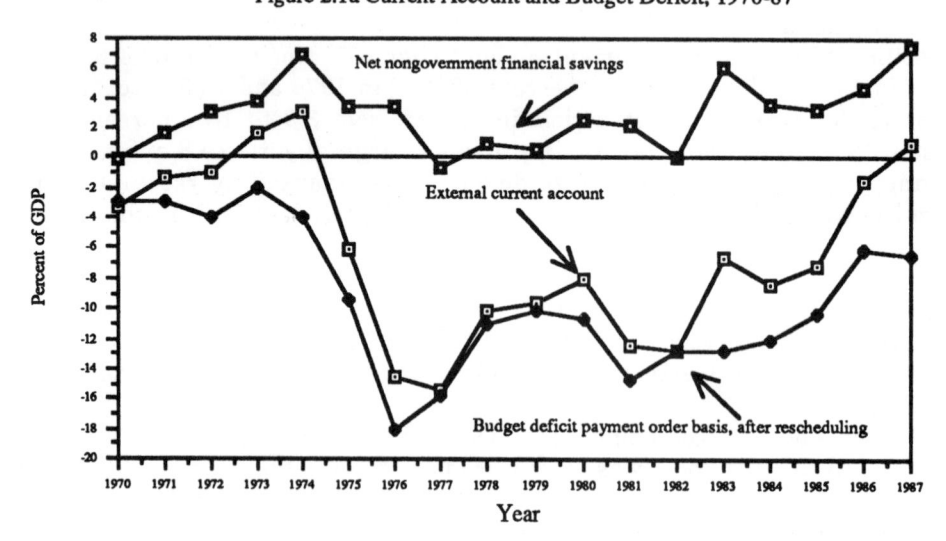

Figure 2.1a Current Account and Budget Deficit, 1970-87

peaking at 130 percent two years earlier. Central Bank reserves had increased to four weeks, a four fold increase over their 1984 level. There was a similar improvement in the net foreign assets of the banking system. Moreover, all of this had been achieved with an inflation rate that never exceeded 15 percent.

The rest of the case study makes extensive use of graphical presentations that are based on the macroeconomic data included in Annex 2, Tables 1 to 13.

As Figure 2.1a and Annex 2, Table 1, show, Morocco's problems began in 1975 with the emergence of a large external current account deficit. From a low of 16.5 percent of GDP in 1977, the deficit improved to less than 1 percent of GDP in 1986 in the wake of export growth, debt rescheduling, import containment, and lower international oil prices. A small surplus emerged in 1987. With the exceptions of 1977 and 1982, savings in the nongovernment sector (S_{ng}) have always exceeded nongovernment investment (I_{ng}). Government expenditures (G), defined on a payment order basis, have always exceeded government tax receipts (T), including dividends from public sector enterprises. To some extent, the external current account has been driven by this budget deficit.[2]

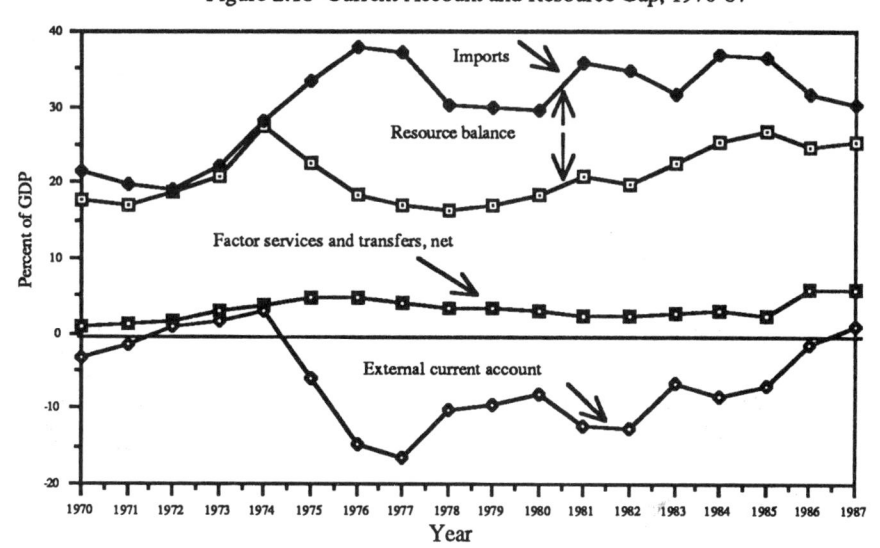

Figure 2.1b Current Account and Resource Gap, 1970-87

---

2. The current account is equal to the sum of the overall Treasury or budget deficit (T-G) and net financial savings ($S_{ng} - I_{ng}$) originating in nongovernmental sectors. It also may be expressed in three other ways: as the sum of the resource gap between exports and imports (X-M) and net factor services and transfers (FST_n) as shown in Figure 2.1b and Annex 2, Table 2; as the sum of domestic savings less investment (S_d-I) and net factor services and transfers (FST_n) as shown in Figure 2.1c and Annex 2, Table 3; and as the difference between national savings (S_n) and investment (I) as shown in Figure 2.1d and Annex 2, Table 4.

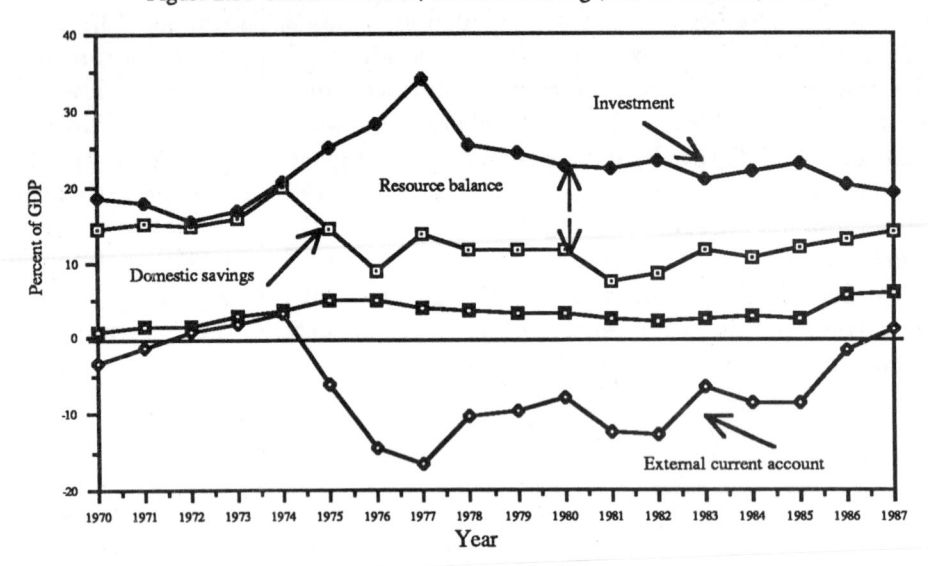

Figure 2.1c Current Account, Domestic Savings, and Investment, 1970-87

Figure 2.1d External Current Account, 1970-87

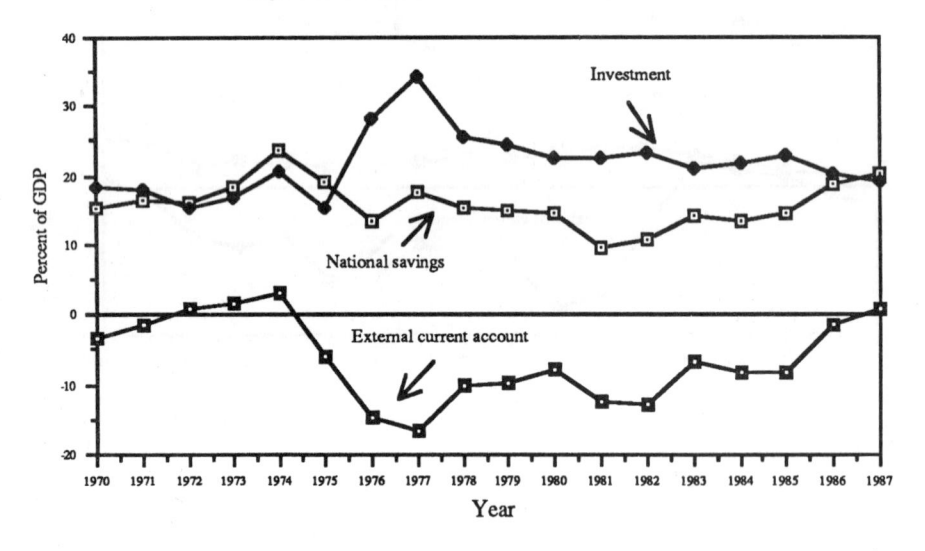

Figures 2.2a and 2.2b—see Annex 2, Table 5—summarize the evolution of government and nongovernment investment from 1970 to 1987. Note from Figure 2.1a, the apparent correlation between government investment and total investment, in the major buildup between 1972 and 1977 as well as in the subsequent decline from 1978 to 1986. From 1980 to 1987, cash outlays on investment by the government were substantially lower than commitments (or investment measured on a payment order basis), so that a good deal of investment was financed by the

accumulation of arrears to suppliers. Nongovernment investment showed a significant upward trend relative to GDP from 1982 to 1986 before retrenching in 1987 (see Figure 2.2a). This is confirmed by Figure 2.2b, which shows the evolution

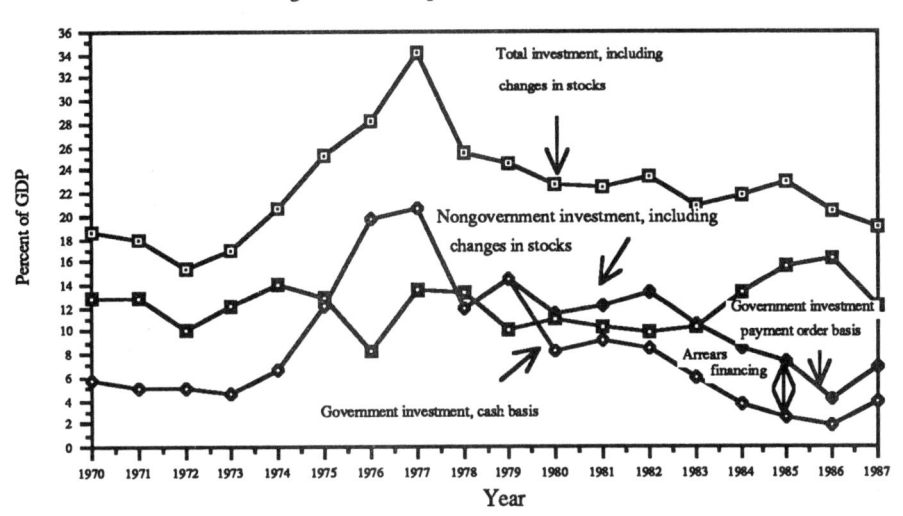

Figure 2.2a Composition of Investment, 1970-87

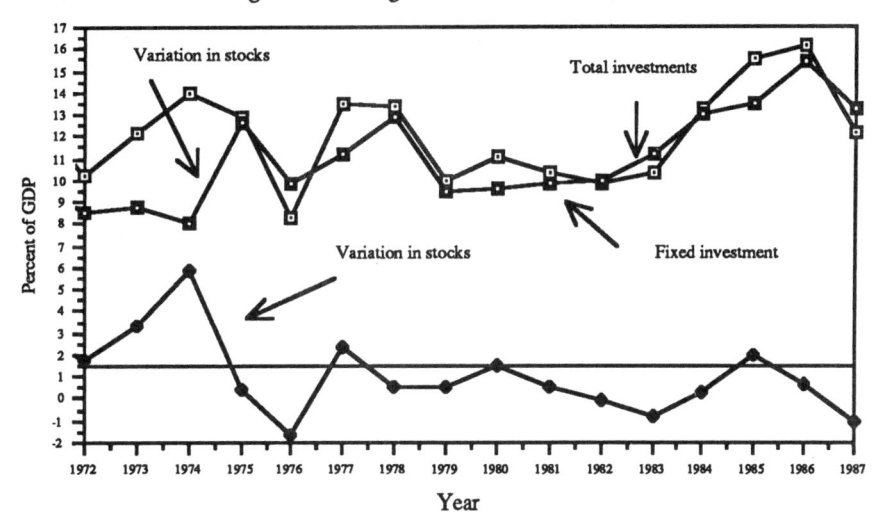

Figure 2.2b Nongovernment Investment, 1972-87

of nongovernment fixed investment (that is, gross investment less changes in stocks).

In turn, Figure 2.3 (Annex 2, Table 6) demonstrates the rapid growth of external debt from about 15 percent of GDP in 1974 to almost 40 percent of GDP in 1977. A three-year period of relative stability followed. By 1985, however, total external debt had risen to between 120 and 130 percent of GDP—most of it held or guaranteed by the government. This accumulation of external debt was accompanied by a rundown of external reserves to a low of two weeks of imports in 1984. As Figure 2.4 (Annex 1, Table 7) shows, there has been some improvement since then—to about 4.5 weeks in 1987. The evolution of net foreign assets, including IMF debt, is depicted in Figure 2.5 (Annex 2, Table 8).

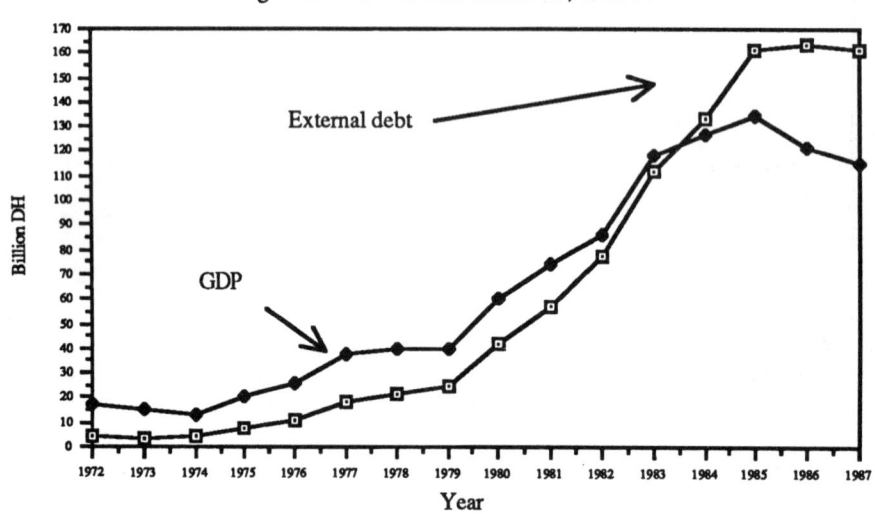

Figure 2.3 GDP and External Debt, 1972-87

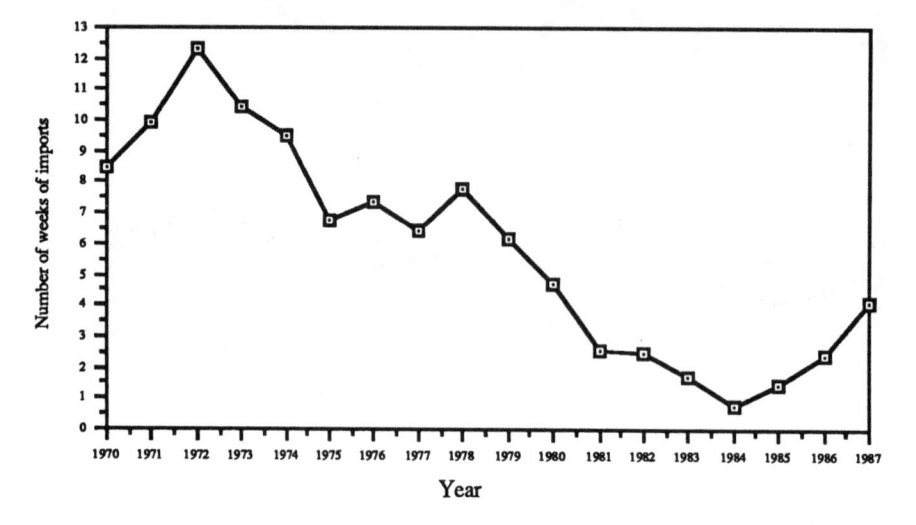

Figure 2.4 Monetary Authority Reserves, 1970-87

Figure 2.6 (Annex 2, Table 9) shows the massive terms of trade effect that occurred in 1974, as well as the movement of the trade weighted exchange rate

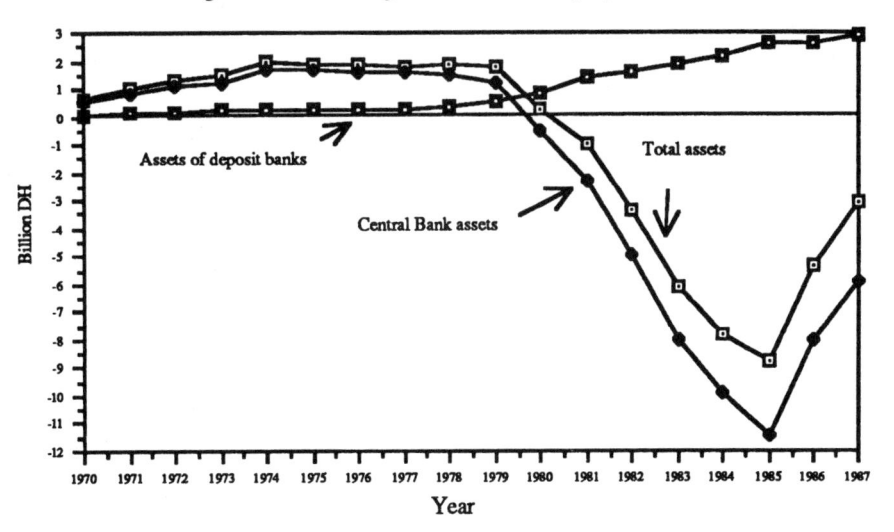

Figure 2.5 Net Foreign Assets of Banking System, 1970-87

between 1967 and 1982 as measured by the nominal and real effective exchange rates. The real exchange rate has been measured using available wholesale price indices as a measure of relative prices in Morocco and abroad. In nominal terms, the dirham appreciated about 10 percent between 1967 and 1974 and then slightly depreciated until 1980. It became somewhat stronger in the next two years. The real exchange rate behaved somewhat differently, exhibiting an 8 percent depreciation

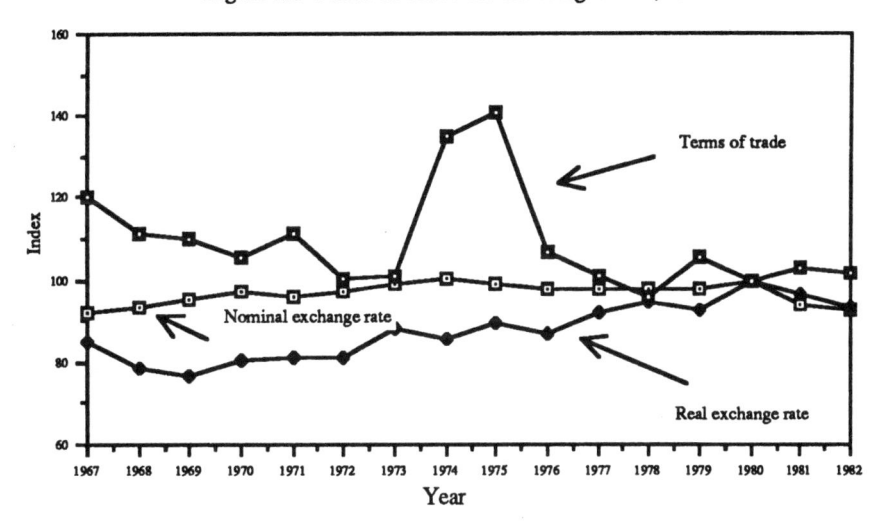

Figure 2.6 Terms of Trade and Exchange Rates, 1967-82

between 1967 and 1969. However, in the following decade there was a steady appreciation of 20 to 25 percent even as the terms of trade showed a net downward trend with the exception of the boom years of 1974 to 1976. A slight depreciation that began in 1980-82 was significantly reinforced after 1983. A more complete picture of exchange rate development is presented in Figure 2.7 (Annex 2, Table 10). Consumer price indices are used to measure relative prices. As can be seen, the consumer price weighted exchange rate has depreciated almost continuously since 1980.

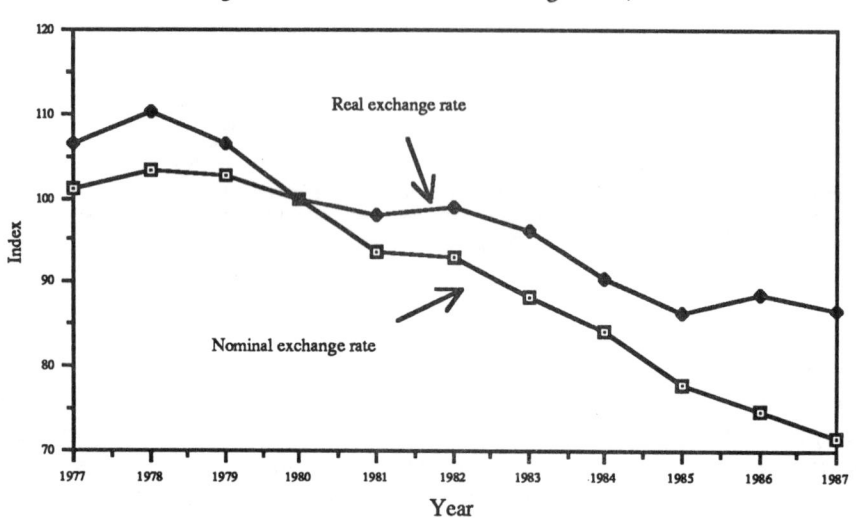

Figure 2.7 Nominal and Real Exchange Rates, 1977-87

Summary of Stabilization and Adjustment Since 1975

From 1975 to 1979, the Moroccan government instituted restrictive credit policies, raised import duties (1975, 1977, 1978, and 1979), tightened quantitative import restrictions (1978 and 1979), and cut public expenditures, particularly on capital goods. By the end of 1979, however, the advent of the second oil shock made it clear that short-term contractionary budget and import policy would not suffice to service public external debt and provide the real imports needed for satisfactory economic growth. A medium-term approach was needed that coupled restrictive demand management and strong supply-side measures.

Accordingly, in 1980, the government borrowed a billion dollars from the IMF under its Extended Fund Facility (EFF). Programs so financed are supposed to combine strong demand and supply-side policies. The EFF loan was to have been accompanied by a structural adjustment loan (SAL) from the World Bank. However, the World Bank's SAL was aborted after inconclusive negotiations. Similarly, the EFF went off track. Its supply-side measures were relatively weak. They concentrated mainly on investment, and for a variety of reasons demand could not be adequately controlled. Consequently, in late 1981, the EFF loan was replaced by a more conventional standby program for 1982. It was predicated on strong demand measures and a desire to continue with the incipient supply-side program.

Following the March 1983 financial crisis, the government again tried, perhaps with greater success, to combine restrictive demand management, especially of public expenditure (in general under IMF auspices), with supply-side adjustment programs financed by the World Bank. Bank programs were designed to promote greater efficiency of resource allocation in the tradables sector with special reference to export promotion.

Government Revenues and Expenditures Including Debt Amortization

The purpose of this section is to discuss in some detail Treasury receipts and expenditures, taking account of debt amortization. In this way, insights will be gained into the political considerations affecting government with respect to stabilization and adjustment as well as the importance of debt rescheduling by foreign creditors. The discussion will draw heavily on Figures 2.8, 2.9, and 2.10 (Annex 2, Tables 11 to 13) that illuminate the main factors influencing the evolution of budgetary receipts, expenditures, and the overall deficit, measured on a payment order or commitment basis since 1970.

Figure 2.8 shows the government's demand for resources, with and without amortization of foreign debt, relative to tax receipts by the government sector. The lowest line (AA) represents total tax and nontax revenues expressed as a percentage of GDP. The next line (BBB) represents total government current and capital expenditure before rescheduling (which became a factor only after 1983) and excluding foreign amortization. Total expenditure after rescheduling is measured by BBB'. The vertical distance between these expenditure lines and the revenue line (AAA) thus constitutes a measure of the overall budget deficit before and after debt relief.

In turn, lines CCC and CCC' represent the government's total demand for resources—that is, the sum of current and capital expenditure plus amortization of foreign loans (principal) before and, starting in 1983, after relief. The difference between these total expenditure lines and revenues, as represented by line AAA, depicts the Treasury's global financing needs including amortization of foreign debt. As Figure 2.8 shows, the overall budget deficit between 1975 and 1977 grew very rapidly. Debt relief, both of interest and principal, was important after 1983 and peaked at 9.4 percent of GDP in 1986.

Figure 2.9 breaks down total expenditures, with and without debt relief, into three components: noninterest current expenditures (RR); capital expenditures (KK); and total debt service (that is, interest plus principal) before and after rescheduling. The impact of rescheduling is shown as the shaded area. This figure highlights the extremely rapid growth of external public debt service and the extent to which this growth has been attenuated by rescheduling. It also shows how the brunt of the reduction in government absorption has been borne by reduced capital formation rather than by noninterest recurrent expenditures. Figure 2.10 breaks down noninterest recurrent expenditures into wages, consumption subsidies and other government transfers, and miscellaneous government expenditures on goods and services. Note the growing importance of consumption subsidies and operating transfers to public enterprises that grew from less than 1 percent of GDP before 1973 to over 6 percent of GDP in 1974.

Figure 2.8 Government Receipts and Expenditures, 1971-87

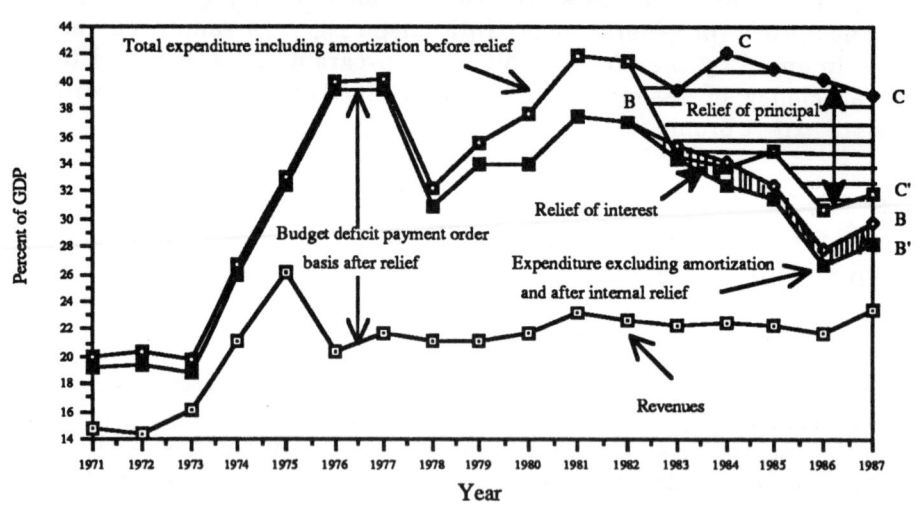

Figure 2.9 Government Expenditure by Category, 1971-87

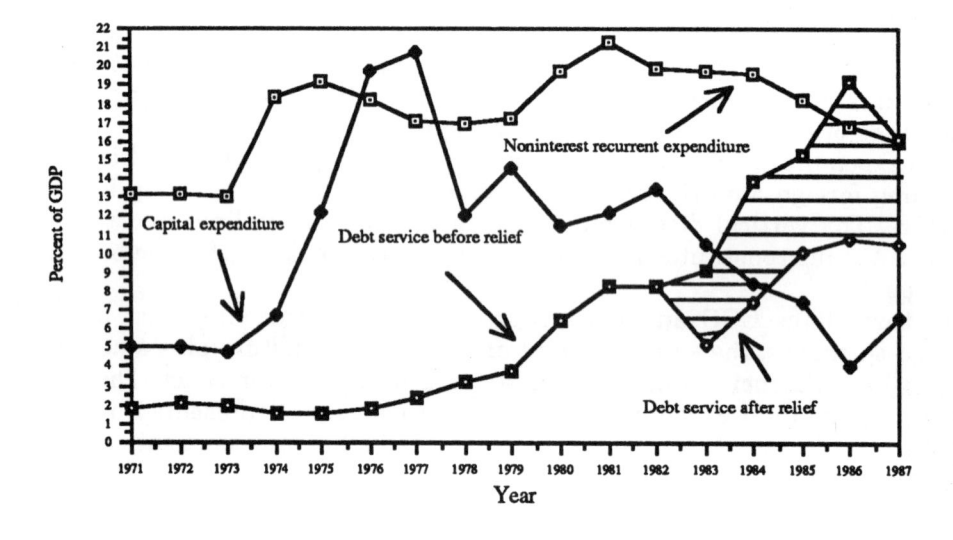

Revenues

The 1973-75 increase in the ratio of total tax and nontax revenues to GDP from 16 to 26 percent was primarily the result of two factors: higher transfers from the state phosphate company (Office Cherifien des Phosphates) attendant upon the boom in phosphate prices and increases in import volumes and prices (for example, crude oil). But, in 1976, overall tax pressure decreased to about 20.7 percent of GDP as the price of rock phosphate diminished to less than its 1974 price in real terms. In the following thirteen years, the ratio fluctuated in the 21 to 24 percent range. The high was 23.2 percent in 1981 and 1987 and the low 21.1 percent in 1986.

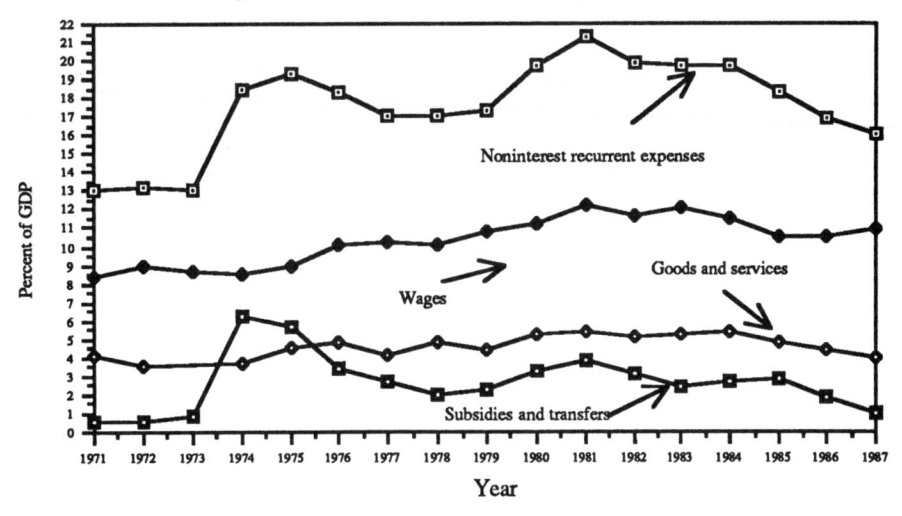

Figure 2.10 Noninterest Recurrent Expenses, 1971-87

Changes in tax policy between 1973 and 1986 reflected the government's revenue and protection concerns and its attempts to rationalize the structure of indirect taxation of imports and domestic provision of goods as well as services. In 1970, Morocco's indirect tax structure had the traditional combination of multiple import levies (equivalent, in effect, to import duties) as well as a multirated manufacturer's sales tax (with a credit mechanism) applied to domestic production and imports. In addition, certain goods (mainly sugar, alcoholic and nonalcoholic beverages, and petroleum products) were subject to excises, whether imported or domestically produced, with no provision for credits.

At the port or point of entry, an import would be subject to a combination of four duties and taxes: the single rate special import tax (SIT), a variable, product-specific import duty (DD), the manufacturer's sales tax (TPS), and finally the single-rate stamp duty (SD). The special import tax and the import duty were levied on the C&F value, and the manufacturer's sales tax on the landed value including the SIT and the DD. However, the stamp duty's base was equal to the sum of the SIT, the DD, and the TPS, so excluding the CIF value of the import duty itself. For example, duties and taxes would be calculated as follows for an import duty with a C&F value (more exactly a customs value) of 1,000 dirham and import duties and taxes as shown:

Tax	Tax rate	Tax base	Receipts	Method of calculation
SIT	15	1000	150	1000*0.15
DD	20	1000	200	1000*0.20
TPS	19	1350	229.50	1350*0.19
SD	10	579.5	57.95	{150+200+229.50}*0.10
Total			637.45	

Between January 1975 and July 1979, the authorities increased import duties and taxes for both protective and revenue reasons. Greater protection was granted in the form of higher customs duties, whereas revenue enhancement was sought by across-the-board increases in the special import tax and the stamp duty as follows:

Date	Rate of special import tax (%)	Rate of stamp duty (%)
Before Dec. 31, 1974	2.5	1
After Oct. 1, 1975	5.0	1
After Jan. 1, 1977	8.0	4
After Jan. 1, 1978	12.0	4
After Aug. 1, 1979	15.0	10

However, on every occasion, no changes were made in the manufacturer's sales tax, which is applied to the larger base of both imports and domestic production of goods and services.

Indirect taxes on imports and domestic goods were not raised again until January 1982, when the standard rate of TPS was increased from 15 to 17 percent. There was a further two-point increase in the standard rate of the TPS in mid-1983.[3]

Thus, in 1982 and 1983, the burden of raising additional revenues shifted from import duties and their equivalents (the SIT and SD) to the TPS. This process went further as the incidence of the SIT was reduced in January 1984 from 15 to 10 percent and a year later to 7.5 percent. These reductions were not accompanied by increases in the rates of the TPS. At the same time, the top rate of the import duty was progressively reduced to 100, 60, and finally 45 percent in January 1986. A third reduction in the SIT from 7.5 to 5 percent was enacted on January 1, 1987, but this was effectively neutralized by an equivalent 2.5 percent increase in most import duties. The manufacturer's sales tax was replaced in April 1986 by a value added tax (VAT) applied to imports as well as domestic manufacturing and wholesale distribution. On average, deductions were increased and rates left unchanged or reduced. This led, at least in the short run, to a downward trend in revenues.

With regard to direct taxes, the most important developments have been in the area of profits taxation. In 1973 as part of its incentive policy, the government introduced a set of sectoral investment codes providing for ten-year partial or total abatements of the profits tax (rates are 40 percent up to 200,000 dirham and 48 percent thereafter) for new industrial firms located outside Casablanca. Similarly, the 1973 export code granted to all firms, wherever located and regardless of their age, a ten-year abatement of the profits tax prorated to the share of export sales in total turnover.

The impact of these exemptions was attenuated to some extent by the introduction in 1980 of the national solidarity tax—Participation à la Solidarité Nationale (PSN)—equal to 10 percent of the theoretical profits tax liability (that is due, but

---

3. The zero and preferential rates of TPS (that is, 4, 6.38, 8, 9, and 12 percent) applied to so-called essential and social goods were left unchanged in both 1982 and 1983.

forgiven by the investment code). In addition, all firms had to constitute an investment reserve of between 5 and 8 percent of their taxable profits. Thus, the cumulative effective rate of profits taxation was equal to about 60 percent for a firm not eligible for any benefits under the investment or export codes. Yet this rate could go as low as 10 to 13 percent for a firm completely exempted from the profits tax under one of the codes.

Beginning in 1983, the investment and export codes were revised. Profit tax abatements and exemptions were reduced in scope, except for export-related profits. Incentives were increased for the use of labor-intensive technologies, via the introduction of employment subsidies, the elimination of accelerated depreciation, and the introduction of import duty exemptions on imported capital goods.

Expenditures

From 1971 to 1973, total government expenditures, including amortization of foreign debt, remained at about 20 percent of GDP (Figure 2.8); in the next few years this ratio approximately doubled. This was primarily the result of the increase in government capital expenditures from 5 to 21 percent of GDP—corresponding to investment in civil infrastructure as well as purchases of military equipment related to hostilities in the former Spanish Sahara (Figure 2.9). The figure also shows a 6 percent upswing in the ratio of noninterest recurrent expenses to GDP between 1971 and 1973.

As Figure 2.10 indicates, the dominant influence was the five-point rise in 1973-74 in consumption subsidies and transfers, as measured by their ratio to GDP. The government decided not to immediately pass on to consumers rises in international prices of sugar, edible oils, butter, milk, petroleum products, and cement. In 1975, however, prices of gasoline were raised to finance the maintenance of subsidies on fuel oil and domestic kerosene. International sugar prices continued their upward spiral during 1975. Sugar-related consumption subsidies rose by more than 150 percent to 2.8 percent of GDP because consumer prices were kept unchanged.

Nevertheless, total expenditures on consumption subsidies, expressed as a proportion of GDP, declined to 5.6 percent of GDP in 1975 and to 2.6 percent in 1977. In retrospect, this seems to have been more attributable to declines in international prices than to increases in domestic prices. Indeed, the maintenance of "low" consumer prices of essential commodities (sugar, edible oils, flour, butter, and milk) was viewed as politically indispensable in order to protect consumers' purchasing power, particularly in poor urban areas. This constraint was perceived by authorities to assume even greater importance in later years.[4]

Expenditures on wages and other goods and services showed smaller increases of 1 to 2 percent of GDP between 1973 and 1977. Similarly, amortization of debt (interest and principal) remained between 1.5 and 2 percent of GDP throughout the period.

As indicated earlier, this large increase in government expenditures was associated with the change in the external current account from a surplus of 2.5 percent of GDP in 1974 to an unsustainable 16 to 18 percent deficit in 1976-77. Because nongovernmental financial savings were insufficient to cover this gap,

---

4. The government was partly motivated by the memory of the disturbances that had occurred in 1965 when sugar prices were raised.

urgent corrective measures were needed to reduce the budget deficit. A start was made in 1977-78 with the following three-pronged policy:

- higher import levies, in particular the special import tax and the stamp tax, on an across-the-board basis for revenue purposes;
- public investment cutbacks;
- tighter quantitative import restrictions as well as increased import duties to provide more protection to domestic industries and conserve foreign exchange.

Relatively little effort was made to cut recurrent expenditures. Also, in 1978, debt service including amortization had risen to 3 percent of GDP, an increase of 50 percent over the preceding year.

The overall impact of these stabilization measures was to cut the government deficit from 18.1 to 11.1 percent of GDP between 1976 and 1978. The external current account situation also improved, with a decline in the deficit from 16.5 to 10.2 percent between 1977 and 1978.[5] Relief was short-lived, however. From 1979 to 1981, the Treasury's total demand for resources, including foreign debt amortization, returned to about 40 percent of GDP because of rising recurrent expenditures on wages, consumption subsidies, and sharply increased debt service (Figure 2.8). Once again government investment was used as a control variable but with much less success than in 1978. In fact, it rose in two out of the three years (Figure 2.9).

Several factors, both external and internal, were at work. The second oil crisis as well as higher prices for wheat, sugar, and vegetable oil forced the government to choose between raising consumer prices and increasing budgetary expenses on consumption subsidies. As in 1974 and 1975, the government's initial reaction was to minimize price increases, which then led to higher budgetary subsidies. Between 1979 and 1980, these rose from 1 to 2 percent of GDP. However, some international prices for commodities continued to increase into 1981. Moreover, the rise of the dollar aggravated the situation.

Again faced with the dilemma of higher consumer prices versus higher budget expenses on subsidies, the government opted, in May 1981, for major increases in the prices of selected basic commodities (flour, sugar, etc.). Following consumer opposition and social unrest, including riots in major urban centers, these increases were partially rescinded. As a result, Treasury expenditure on consumer subsidies for 1981 attained 2.8 percent of GDP, their highest level since 1976.

This upward trend in recurrent expenditure was reinforced by increases in outlays on goods/services of 13 percent and on wages of 30 percent—12.2 and 8.4 percent of GDP, respectively. As a result of these developments, total noninterest recurrent expenditures rose from 17 to 21 percent of GDP. The local currency cost of Treasury debt service advanced from 3.2 to 8.3 percent of GDP between 1978 and 1981. This resulted partly from the intrinsic profile of debt repayments and partly from the upswing in international interest rates as well as the dollar. Government investment rebounded from its 1978 low of 12 percent to 14.5 percent of GDP one year

---

5. Despite these improvements, the Treasury's total demand for resources, including amortization, still vastly exceeded revenues by over 11 percent. The resulting financing gap was filled by further borrowing from domestic and foreign sources, both commercial and bilateral. Domestic borrowing was primarily from commercial banks that were compelled to invest 30 percent of their sight deposits in low interest Treasury securities. However, the Treasury also benefited from substantial advances from the Central Bank.

later. After a temporary relapse to 11 percent in 1980, this ratio recovered in 1981 and 1982 with the advent of the 1981-85 Economic Plan.

Thus, between 1978 and 1981, total government demand, inclusive of debt amortization, systematically increased from 32.2 percent to 41 percent of GDP, slightly higher than the previous 1977 peak of 40 percent. In the same period, the ratio of revenues to GDP had increased from 21.2 to 23.2 percent of GDP.[6] Consequently, the Treasury's gross financing requirements rose from 11 to 18 percent of GDP. Furthermore, after 1980, Treasury revenues were less than current expenditures so that not only debt amortization but also public investment had to be financed with borrowed funds that were procured both domestically and abroad.

Especially in 1980 and 1981, there were important changes in the sources of this financing. In the first place, there was a sharp decline in loans contracted in the international financial market and a corresponding increase in resources raised from bilateral sources, in particular from Saudi Arabia. At the same time, during the latter part of 1980, a loan of SDR 800 million (that is, approximately US$1.05 billion) was obtained from the IMF under its Extended Fund Facility. This loan was supposed to be disbursed over a three-year period. However, inability to meet the targets in 1981 led to the loan's cancellation at the end of the year. The advent and demise of this attempt to adjust show the extent to which the government's freedom of action in the policy area was beginning to be constrained by the exigencies of its external debtors. Again, this was a foretaste of things to come.

The year 1982 was relatively stable in the sense that total public demand for resources, inclusive of amortization, declined marginally from the level of 41 percent of GDP in 1981. There was a downward trend in all categories of recurrent expenditure, but the investment to GDP ratio rose along with continuing implementation of the 1981-85 plan. However, since revenues amounted to only 22.7 percent of GDP,[7] the Treasury's gross financing needs represented 18.2 percent of GDP. This gap was fully financed primarily from bilateral donors (in particular Saudi Arabia), the international financial markets, and the IMF, which had replaced the EFF with a one-year standby of an equivalent amount.

Notwithstanding this apparent stability, during 1982 it became increasingly clear that serious debt servicing problems lay ahead. Yet on the domestic front, the 1981-85 Economic Plan called for a further expansion in public investment in 1983. In addition, the decline in consumer purchasing power of the preceding years was perceived as a potential source of social difficulties.

In response to these pressures, the government passed an expansionary budget for 1983. But by March, foreign exchange reserves had been virtually depleted, and new loans could not be quickly contracted. Consequently, the government was forced to implement emergency import restrictions in the form of administrative licensing of all imports. It also requested debt rescheduling and new funds from its private and public creditors. Furthermore, it became clear that debt relief would be necessary on a multiyear basis in view of the rapidly increasing amortization obligations. Management of the public finance situation became considerably more difficult after 1983, again because of internal and external factors.

The number of lenders actively interested in economic management widened as the government sought relief from bilateral lenders and foreign commercial

---

6. This was mainly because of a mini boom in phosphate prices in 1981 that had resulted in much larger transfers from OCP to the Treasury.

7. There had been a two-point increase in the standard rate of the manufacturer's sales tax in effect since January 1, 1982.

banks. The World Bank began to provide substantial balance of payments support that was conditional, however, upon the adoption of policy reforms in the industrial, financial, agricultural, and public enterprise sectors. The first two Bank loans— ITPA1 and 2—were designed to reform the structure of incentives and relative prices of exportables, importables, and nontraded goods. The goal was to increase the domestic supply of foreign exchange through economically efficient exports and import substitution, hence improving the efficiency of resource allocation. To this end, key proposals were a devaluation, accompanied by a compensating reduction of the import levies, especially the SIT and import duties; a reduction in protective import duties (specifically lowering the maximum); and a liberalization of competing imports. The revenue shortfall was to be recovered by an increase in the rates of the TPS and/or a reduction in government expenditures. Furthermore, policy reforms were proposed for the financial, agricultural, and public sectors.

From the government's standpoint, these reforms had to be integrated into a program of action that was heavily influenced by the exigencies of the IMF's existing and future standbys. Although the World Bank and IMF programs had many points in common, the government felt that they were fundamentally contradictory on one point—namely, the reduction of the import levies, especially the SIT. This proved to be a major stumbling block since the IMF's programs continued to emphasize reductions in the Treasury deficit. Also, the government could not afford to ignore the wishes of the international financial community. The willingness of the banks to reschedule, or not, became a critical influence on the overall supply of foreign exchange.

One domestic constraint on government behavior was the perceived need to avoid increases in consumer prices of basic commodities in the interest of avoiding social disruption. Another was the national goal of protecting the former Spanish Sahara. Finally, of course, the economy's future growth was argued to be contingent upon the maintenance of a minimum amount of investment.

Figure 2.8 indicates the extent to which rescheduling of principal and interest has provided debt relief. In its absence, gross Treasury demand for resources would have been between 38 and 40 percent of GDP every year since 1983 (see line CCC). Relief has lowered this to between 30 and 32 percent. However, there has been a substantial reduction in total government expenditure relative to GDP, of which only a part (the shaded area between BBB and BBB') consists of interest rescheduling. The major burden of expenditure reduction continued to fall on public investment, which was reduced between 1982 and 1986 from 13 to about 4 percent of GDP, its lowest level since 1970. Reduction in noninterest current expenditure was much less pronounced, from about 21 to 16 percent of GDP between 1982 and 1987 as a result of a decline in all categories of expenditure relative to GDP.

Conclusion

In the analysis of stabilization and adjustment in Morocco from 1975 to 1987, 1983 is a natural dividing point. That year the external payments crisis began, and the government was compelled to request emergency financial assistance from external creditors. Moreover, 1983 marked a significant turning point in economic policy. Sectoral adjustment policies, combined with ongoing stabilization programs, sought to ameliorate resource allocation in the interest of economically efficient growth. Before 1983, economic policy had tended to emphasize stabilization rather than adjustment.

Several macroeconomic factors, both external and internal, impinged simultaneously upon Morocco's economy. Primary external factors included drought and terms of trade changes, due mainly to variations in the prices of phosphates and crude oil. The main domestic factors were the expansion of public investment, particularly in 1973 and 1977 and again in 1981-82, as well as a substantial increase in recurrent expenditures related to national defense, the preservation of purchasing power, and other deserving causes such as health and education.

This volatile external environment coupled with excess domestic expenditure led to unsustainably large external current account and Treasury deficits that were financed by a buildup of external debt. The government's initial reaction in the mid-1970s was to enact stabilization policies built around tight monetary policy, higher import duties, and more restrictive import policies. Some attempts also were made to control expenditures. However, there has been a reluctance to decrease consumer subsidies because of the political sensitivity of raising consumer prices of certain basic commodities. In this regard, experience from the 1960s had a cautionary value. When sugar prices were raised in 1965, riots followed.

By 1983, the microeconomic disadvantages of the stabilization policies of the 1970s had become apparent—namely, that they tended to reinforce import substitution and increase anti-export bias resulting from higher tariffs and tighter quantitative restrictions. Consequently, post-1983 policies attempted to combine stabilization policies with adjustment policies that avoided, indeed reversed, the undesirable effects of the early approach.

3
Methodology

This chapter presents a methodology for addressing the macroeconomic and microeconomic aspects of different approaches to stabilization and adjustment. The starting point is to set forth the key macroeconomic accounting relationships and behavioral equations that relate macroeconomic variables such as exports, imports, tax receipts, and government expenditures to wages, prices, interest rates, and the exchange rate (that is, the price of foreign exchange).[1] Then methodologies are described for measuring (a) production incentives for exports as well as the domestic market and (b) economic profitability of activities. These microeconomic measures are placed in a broad macroeconomic framework. A short discussion of the relevant aspects of tax policy follows. The chapter concludes by touching upon aspects of interest rate policy and management relevant to the design of stabilization and adjustment programs.

A Framework for Comparing Approaches to Adjustment and Stabilization

Most discussions of stabilization make use of various decompositions of the current account deficit into its component parts. Such is the approach taken here. The current account gap can be expressed as follows:

$$(3\text{-}1) \quad CA = S_n - I$$
$$(3\text{-}2) \quad CA = [X - M] + FS_n + FT_n$$
$$(3\text{-}3) \quad CA = [S_d - I] + FS_n + FT_n$$
$$(3\text{-}4) \quad CA = [S_{ng} - I_{ng}] + [S_g - I_g]$$
$$(3\text{-}5) \quad CA = [S_{ng} - I_{ng}] + [T - G]$$

where

---

1. Discussion of policy options between the different ministries, as well as between the government and external creditors (such as the IMF and the World Bank), may be enhanced by such a global framework encompassing both micro- and macroeconomic dimensions.

CA = current account gap
S_n = national savings
I = investment
S_d = domestic savings
X = exports of goods and nonfactor services
M = imports of goods and nonfactor services
FS_n = net factor service payments
FT_n = net foreign transfers
S_{ng} = nongovernment saving, originating in productive sectors and households
I_{ng} = nongovernment investment
S_g = government (budgetary) savings
I_g = government investment
T = government receipts
G = total government expenditures on goods and services including interest, but excluding amortization of principal.

As classified below, alternative formulations of the current account deficit highlight two approaches to the stabilization and adjustment process:

Approach 1

$$CA = [X - M] + FS_n + FT_n$$
$$CA = [S_d - I] + FS_n + FT_n = [S_d + FS_n + FT_n] - I$$
$$CA = [S_n - I]$$

Approach 2

$$CA = [T - G] + [S_{ng} - I_{ng}]$$
$$CA = [S_g - I_g] + [S_{ng} - I_{ng}]$$

The first approach may be used to examine the evolution of (a) exports and imports of goods and nonfactor services—that is, the economy's resource gap (X - M); (b) net factor service payments (FS_n)—that is, worker remittances and net investment income; and (c) other net foreign transfers (FT_n). Because the resource gap is equal to the difference between domestic savings and investment, the first approach also may be used to highlight the shortfall of domestic savings relative to investment as an element in the country's current account. This approach shows the performance of exports, imports, fixed capital formation, worker remittances, and debt service without distinguishing between government and nongovernment resource flows.

The second approach defines the current account gap as the sum of the budget deficit and nongovernment savings investment gaps. It does not explicitly mention exports and imports, but emphasizes instead the decomposition of the current account gap (or the national savings investment gap) into its sectoral components. This approach has the advantage of correlating the current account gap and the savings investment gap (or equivalently the income expenditure gap) of the government and the nongovernment sectors.

Despite their difference in emphasis, the two approaches are closely linked. This can be seen when they are put in the context of causal relationships explaining the level of each variable. From standard national income accounts the gross national product may be defined as follows:

(3-6)　　GNP = GDP (M) + FS_n + FT_n
(3-7)　　GNP = C + I + G + X(M) + FS_n + FT_n - M,

where GDP denotes gross domestic product.

From the above, one may obtain

(3-8)　　$[S_{ng} - I_{ng}] + [S_g - I_g]$ = X (M) - M + FS_n + FT_n
(3-9)　　$[S_{ng} - I_{ng}] + [S_g - I_g]$ = CA
(3-10)　　$[S_{ng} - I_{ng}] + [S_g - I_g]$ = ΔNFA
(3-11)　　$[S_{ng} - I_{ng}] - [S_g - I_g]$ = ΔNFA_g + ΔNFA_p
(3-12)　　ΔM_2 = ΔNFA + ΔCR_g + ΔCR_{ng}

Equation 3-8 encapsulates the two approaches to the current account. In turn, equations 3-10 and 3-11 indicate that the current account gap equals the variation (denoted by Δ) in net foreign assets (NFA), decomposed into its government and nongovernment components (NFA_g and ΔNFA_{ng}, respectively). Equation 3-12 explains the link between the money supply (M_2), net foreign assets, and bank credit (CR) to the Treasury and the nongovernment sector.

GDP and exports depend upon the level of imports (M). More exactly, exports depend upon directly used imports (M_x) and also on the level of imports used by the nonexporting sector (M_{nx}), excluding final consumption goods (M_c). The nonexporting sector is assumed to provide goods and services to the exporting sector. Furthermore, it may be shown that

(3-13)　　$S_g - I_g$ = T - G
(3-14)　　$S_g - I_g$ = ΔNFA_g - ΔCR_g
(3-15)　　$S_{ng} - I_{ng}$ = ΔNFA_{ng} + ΔCR_g
(3-16)　　$S_{ng} - I_{ng}$ = ΔM_2 - ΔNFA_g - ΔCR_{ng}.

These equations express the relationship between the real and monetary accounting of the private and government savings investment gaps.

It is necessary to specify how taxes (T) are levied. Imports used by exporters (M_x) are assumed not to pay any import duties or taxes in *effective* terms.[2] Other imports (M_{nx} and M_c) bear import duties at an average rate of v_d and an import turnover tax of v_m. Domestic sales pay a turnover tax (v_d), but the import turnover tax on certain imports—generally raw materials (M_{nx})—may be credited against this. Factor incomes—wages and profits—are directly taxed at a combined average

---

2. This is an assumption for expository purposes only. In practice, imported inputs used in exports do pay duties and taxes that may be more or less effectively suspended or restituted. If this price is imperfect, the *effective* rate of taxation would, of course, be positive. In practice, attaining the ideal of zero effective taxation is an important objective of structural adjustment programs.

rate of t_d for firms selling to the domestic market and t_x for exporters.[3] Finally, exports may bear export duties or taxes at an average rate of d_x. Under these assumptions, tax collections may be represented as follows:

(3-17) $\quad T = d_m (M_{nx} + M_c) \quad$ | import duties
$\quad + $
$\quad v_m (1 + d_m) (M_{nx} + M_c) \quad$ | import turnover tax
$\quad + $
$\quad v_d S_d - v_m (1 + d_m) (M_{nx}) \quad$ | domestic turnover tax less import turnover tax on raw materials (M_{nx})
$\quad + $
$\quad t_d\, VA_d \quad$ | direct taxes on wages and profits of firms selling to the domestic market, expressed relative to value added
$\quad + $
$\quad t_x\, VA_x \quad$ | direct taxes on wages and profits of exporters, expressed relative to value added
$\quad + $
$\quad d_x X \quad$ | export duties and taxes levied by customs

Similarly, total government expenditure may be written

(3-18) $\quad G = G(T, B_g) = C_{gs} + C_s + I_g + DS_i$

where B_g denotes total government borrowing, C_{gs} expenditures on goods and services, C_s consumption subsidies, I_g capital formation, and DS_i interest payments on public debt.

Similarly, nongovernment savings is postulated to be a function of GDP and worker remittances (R):

(3-19) $\quad S_{ng} = S_{ng} (GDP, R).$

The export supply function can be expressed

(3-20) $\quad X = X [rp_x (1 - d_x)/r\, p_m (1 + d), M_{nx}, w/rp_x]$

where r is the price of foreign exchange, p_x and p_m are the world prices of exports and import substitutes in foreign exchange, and w is the wage rate. Thus, exports depend on the relative price, in domestic terms, between exports and import substitutes, the availability of foreign exchange to the nonexport sector, and the real wage in terms of exports.

Wages are assumed to be related to domestic prices so that

---

3. It is assumed that t_d and t_x are unequal to highlight the fact that in some countries, including Morocco, some exporting firms may obtain a preferential rate of direct taxation of profits and hence their value added. It is not true, of course, that all exported value added is generated by 100 percent export-oriented firms.

$$(3\text{-}21) \quad w = \Sigma (1 + v_i) Q_i p_i$$

where v_i is the effective rate of consumption taxation of the i-th good, net of consumption subsidies.

Imports of goods are determined as follows:

$$(3\text{-}22) \quad M_x = M[X, r, p_x, p_m]$$
$$(3\text{-}23) \quad M_{nx} + M_c = M[rp_m (1 + d)(1 + v_m), GDP, R, B_{ng}]$$

where B_{ng} is nongovernment domestic and foreign borrowing.

Finally, foreign remittances (R) and interest payments on debt service (DS_i)—the main components of the factor service balance FS_n—are assumed to be related to interest rates at home (i_d) and abroad (i_f):

$$(3\text{-}24) \quad R = R(i_d, i_f)$$
$$(3\text{-}25) \quad DS_i = r[i_f FB(-1) - DR]$$

where DS_i denotes the domestic currency cost of the interest component of debt service on outstanding and disbursed debt at the start of the period—i_f FB (-1)—less rescheduling of interest (DR).

This model can be used for comparing different policy options for growth, stabilization, and adjustment and their impact on external and internal imbalances. It is also useful in discussing the supply of foreign exchange from exports (X) and remittances (R) as well as present and future foreign borrowing (FB). For example, a purely budgetary approach or stabilization approach to reducing a current account deficit might seek to reduce the overall Treasury deficit through expenditure reduction and raising import duties that would reinforce anti-export bias.[4]

A more adjustment-oriented approach to the same problem would recognize the importance of macroeconomic—especially budgetary—stabilization, but also would stress the need for a package of measures, budgetary and nonbudgetary alike, that seeks to avoid an increase in anti-export bias and if possible to reverse it. On the budgetary side, this would imply the use of higher internal, trade neutral, indirect and direct taxes, rather than trade-distorting import duties to increase government revenues.[5] In addition, there would be reductions in expenditure, yet these would strive to spare infrastructure and services that enhance the efficiency of international trade.

The adjustment approach therefore emphasizes increasing the supply of efficiently produced tradable goods, particularly exports. In that connection, key measures would include exchange rate adjustment to recover past appreciation relative to principal competitors, phased removal of nontariff barriers to imports, and tariff reduction to low and quasi uniform levels in order to encourage domestic

---

4. The increase in anti-export bias would result from (a) an increase in the relative domestic price of import substitutes and exports as well as (b) a correlative overvaluation of the price of foreign exchange.

5. Thus, optimal tax policy would be based, as far as possible, on the adoption of pure consumption taxes levied on both imports and domestic transactions as the primary source of commodity taxation. Similarly, trade neutral direct taxation would be encouraged.

firms to competitive efficiency in terms of price, quality, and delivery time relative to international markets.

Measurement of Production Incentives

To measure production incentives for domestic sales and exports, as well as financial and economic profitability, the case study uses the effective protection methodology based on firm-level data. For any product sold on the domestic or export market, this methodology compares (a) domestic and international prices (expressed in terms of nominal protection coefficients) and (b) unit value added margins in domestic and international prices (expressed in terms of effective protection and subsidy coefficients). These indicators may then be used to discuss incentives to produce for the domestic and export markets. In the former case, a distinction is drawn between different categories of goods (food, final consumption, intermediate goods, and capital equipment) that may have different incentive structures.

Specifically, the *nominal protection coefficient* (NPC) is defined as the ratio, in domestic prices, of the ex factory producer price (p_d)—net of sales tax—to the border price (p_w) of the same or equivalent good, using the import or export price as appropriate. Similarly, for an input, the NPC is the ratio of its cost to the user in domestic prices and the same good in international prices. The calculation of NPCs therefore requires careful domestic and international price comparisons.

The NPC, however, is not a good measure of the real or effective protection granted to the production process since manufacturers are interested in enlarging their unit transformation or value added margins. In this regard, protection is better measured by the *effective protection coefficient* (EPC), which is defined as the ratio of the value added incorporated in a product, calculated first in domestic prices and then in international prices:

(3-26) $EPC = VA_{pd}/rVA_{pw}$
(3-27) $EPC = [p_{j,d} - \Sigma a_i p_{ij,d}]/r[p_{j,w} - \Sigma a_{ij}p_{i,w}]$

where $p_{j,d}$ and $p_{j,w}$ are prices of the j-th good in domestic and world prices, a_{ij} denotes input-output coefficients, and r is the price of foreign exchange.

The EPC thus captures the combined effect on firms' unit value added margins of the nominal protection granted to the finished good and its inputs, both imported and of domestic origin. Production incentives also can be granted in the form of nontrade tax or financial subsidies, which are captured by the *effective subsidy coefficient* (ESC). The ESC is calculated by adding to the numerator of the EPC the net value of any such subsidies and levies.

An NPC > 1 for an output indicates that the firm's output price is greater than the world price. Likewise, for an input an NPC > 1 arises when a manufacturer pays, on account of import duties and taxes or protection granted to its domestic suppliers, more than international prices for its inputs. Analogous definitions hold for NPCs that are less than unity.

In the same vein, EPCs greater than unity indicate positive effective protection to processing because unit transformation or value added margins in domestic prices exceed their values in international prices. EPCs of less than unity imply negative effective protection.

These nominal and effective protection coefficients may be used to compare the structure of incentives for domestic sale and export. For example, if the EPC for the domestic market is greater than for exports, there is a relative anti-export bias. If, as well, the export EPC is less than unity, a situation of absolute anti-export bias exists.

The preceding indicators are generally calculated at the prevailing exchange rate. It is customary, however, to make an adjustment for any exchange rate overvaluation relative to its equilibrium value by dividing the indicators by a coefficient $(1 + x)$ where x is the percentage overvaluation of the exchange rate. Such adjusted NPCs, EPCs, and ESCs are referred to as net nominal and effective protection and effective subsidy coefficients. The equilibrium exchange rate is that which would prevail in the absence of all tariff and nontariff restrictions on imports and exports (except as justified on monopoly grounds). Its estimation would require a full model of the economy. To measure economic profitability, the case study makes use of the familiar domestic resource cost (DRC) of foreign exchange, which can be expressed

(3-28) $\quad DRC = [p_k K + wL]/rVA_{pw}$

where the numerator is the shadow factor cost of earning a unit of net foreign exchange. An activity is said to be economically profitable if its DRC is less than unity. Incentive analyses carefully compare the DRC and the EPCs for the same activities to assess the correlation between economic profitability and the incentives received in the form of effective protection or subsidy.[6]

Although the EPC methodology is essentially microeconomic, it can be helpful in macroeconomic analysis because nominal protection coefficients, effective protection coefficients, and domestic resource costs depend on the exchange rate, import duties, and internal indirect taxes. Consequently, these indicators may be used to compare various policy options (exchange rate changes, import tariffs, and export subsidies) for modifying the structure of incentives in general and reducing anti-export bias in particular.[7]

Analysis of exchange rate adjustment is especially important because the EPC methodology explicitly tries to adjust overvaluation relative to the equilibrium exchange rate. Most macroeconomic analyses tend to analyze the consequences only of exchange rate appreciation or depreciation in purchasing power parity (PPP) since that is relative to a base year. There is, however, no reason why the two

---

6. An alternative would be to discuss economic profitability in terms of the gross resource cost of foreign exchange (GRCFE). A variant of the DRC, it can be expressed

$\quad GRCFE = [rM + p_k K + WL]/rQ$

where Q is the gross value of foreign exchange earned or saved by the activity, and M is the foreign exchange cost of imported raw materials and intermediate goods. An activity is economically profitable if GRCFE is greater than unity.

In this sense, there is not a great deal of difference between a DRC and a GRCFE. But in relating measures of economic profitability and incentive indicators, the GRCFE and the NPC would be compared, whereas DRCs are compared with EPCs. The GRCFE indicates a particular product's profitability in terms of its shadow cost of production relative to world prices. The difference between the GRCFE and unity may be thought of as the minimum import duty or export subsidy that the activity would require if factors were to earn their economic opportunity costs (assuming that all inputs were purchased at world or shadow prices).

7. Strictly, such an analysis would require a general equilibrium model. One was not constructed in the Moroccan case, but incentives analysis of a well-chosen sample of activities can be a useful starting point.

indicators should give the same estimation of overvaluation. This is important since analysis of exchange rate movements and misalignment by the IMF has traditionally relied only on PPP measures. These measures do not take into account the possibility that the exchange rate in the base year was significantly overvalued relative to its equilibrium value, as defined earlier.

Tax Policy

Adjustment analysis also should focus on the design of optimal tax policy, particularly as it relates to international trade and factor income (wages and profits). The tax equation presented earlier (3-17) can be of great assistance in this regard.

Insofar as trade taxes are concerned, it is better to forgo raising revenues through import duties (d_m). Instead, recourse should be had to noncascading internal indirect taxes—that is, a consumption tax of the sales or value added variety. These should be levied at equal rates on imports and domestic transactions, including, if possible, wholesale distribution. Thus, in terms of equation (3-17), v_m and v_d should be equal.

The reasons for this are well known. Raising revenues through import duties introduces distortions in the structure of incentives. This tends to increase anti-export bias in the form of an overvalued exchange rate, compared with its equilibrium value, as well as higher effective protection coefficients for the domestic market compared with exports. These difficulties may be attenuated if import duties, levied for revenue purposes, are replaced by internal indirect taxes.[8]

Export taxes are to be eschewed except for commodities in which the country may have market power. In terms of equation (3-17), d_x should be zero except for goods when the country may have monopoly power in trade.

On the direct tax side, two points should be remembered. First, the structure of personal income taxation should be examined with a view to lowering unduly high marginal rates, which may undesirably affect wage costs, particularly of domestic and foreign skilled labor.[9] Second, the structure and levels of corporate profits taxation, as well as the revenue costs of exemptions given to various activities or firms, should be examined. Special attention should be given to the relative taxation of traded and nontraded activities (such as construction) with a view to eliminating preferential treatment of the latter compared with the former. Ideally, corporate profits from tradable activities should be taxed at the same rate regardless of whether they originate from domestic sales or exports. In other words, direct taxation should be trade neutral. If preferential rates for exports are adopted, they should be minimized and complete exemptions eschewed. This is particularly true in economics where exports are to be a significant source of growth since associated revenue loss of "big" exemptions may be significant.

Interest Rate Policy

Another methodological consideration pertains to interest rate policy, where there are also significant micro and macro links. Interest rates are important to

---

8. Of course, anti-export bias and protection of the domestic market may result from administrative or quantitative restriction of imports.
9. Such high taxes will increase economic costs as measured by the domestic or gross resource cost of foreign exchange, thus adversely affecting competitiveness.

producers and the government alike. They can affect key economic variables such as savings, financial savings (that is, the allocation of real savings between real and financial assets), and the cost of loanable funds for investment and working capital. To design stabilization and adjustment programs, a clear understanding of these considerations and the relation between them is needed.

Opinions differ concerning the effect of interest rates on the level and composition of real savings. For some, the key impact of interest rates is to raise the supply of savings; for others, savings respond mainly to income levels, while interest rates affect primarily the allocation of savings between real and financial assets. On the demand side, interest rates may affect not only the level of investment but also the demand for loanable funds. In both cases, higher interest rates are likely to restrain import demand for capital goods. Government authorities are particularly sensitive on this point, as they are to the upward impact of increasing interest rates on financing Treasury debt.

Governments may use interest rate policy to reduce the cost of financing for a wide variety of activities such as Treasury operations, agriculture, exports, and certain categories of investment. Special refinancing windows at the Central Bank may then have to be opened. It may be necessary to introduce obligatory placement requirements on commercial bank deposits in order to channel the financial resources to the desired borrowers.[10]

These devices represent, of course, an implicit transfer to the beneficiary, but they are by no means costless from an economic standpoint. For example, channeling financial resources to certain borrowers at preferential interest rates entails a reduction in bank revenues. This may then put downward pressures on deposit rates with an attendant negative impact on deposit mobilization. In a similar vein, obligatory placement requirements reduce bank spreads and may even make them negative for long-term deposits. Under these circumstances, banks have no interest in accepting long-term deposits even if their mobilization is an explicit objective of government policy. Finally, the access to such preferential interest rates may have to be rationed in productive sectors, in order to contain their cost to financial intermediaries and the budget. Such practices may not ensure that the financial resources are channeled to the best use from an economic standpoint.

Preferential interest rates affect the structure of incentives by reducing the cost of financing transactions. Unit value added or processing margins may then be increased. Incentives analysis generally takes account of such incentives by incorporating them into the effective subsidy coefficient (ESC), as defined earlier.

---

10. Obligatory placement requirements compel banks to invest a certain proportion of their deposits in obligations issued by the Treasury or specialized lending agencies such as agricultural and development banks.

4
Economic Policy Reform, 1973-82

The purpose of this chapter is to explore in greater detail the first phase of Morocco's attempt to stabilize and adjust. In fact, the period 1973-82 may be conveniently subdivided into two subperiods: 1973-79 and 1980-82. In both periods, resources were obtained from various IMF facilities (The Gold Fund, First Credit Tranche, Compensatory Financing Facility, the Extended Fund Facility, and Standby Facilities). Also, in 1980 and 1981, the World Bank attempted, unsuccessfully, to put together a structural adjustment loan (SAL) to complement the macroeconomic foundation laid by the IMF's Extended Fund Facility (EFF). To some extent, both the Fund and the Bank considered their resources as providing part of the estimated $US2 billion of external financing required by the 1981-85 Economic Plan. The chapter concludes with a preliminary comparison of the different approaches to stabilization and adjustment.

Expansion and Stabilization, 1973-79

The 1973-77 Economic Plan

The 1973-77 Economic Plan, the Moroccan government decided, would be more ambitious than previous plans. Its objectives were to promote the growth of exports, mainly phosphates and their derivatives (phosphoric acid and fertilizers), as well as public and private import substituting industries. To this end, new investment and export investment codes were adopted in August 1973. The plan was substantially revised in 1974 in the light of the increase in phosphate prices in the world market, which more than outweighed the negative impact of the 1973 oil shock. The trade balance, as measured by the resource gap, was in rough equilibrium in 1974, but positive net earnings from factor services and current transfers, especially worker remittances, resulted in a current account surplus of about 3.1 percent of GDP (see Figures 2.1a and 2.1b).

The revision of the plan led to a sharp jump in public investment beginning in 1975, which the government financed partly by foreign borrowing and raising import levies, especially the special import tax (SIT), which increased from 2.5 to 5 percent (in January 1975). In addition, the current account surplus gave way in 1975 to a deficit equal to 6.1 percent of GDP. The Treasury deficit increased from 3.9 to

9.5 percent of GDP. Expansion continued into 1976 as public investment attained 28.1 percent of GDP, despite the collapse of the world phosphate market and poor performance in nonphosphate exports as well as higher imports as a result of poor rainfall and a relaxation of import controls. At the same time, recurrent public expenditures were rising on account of military outlays and higher consumption subsidies on essential commodities (sugar, edible oils, and petroleum products) whose consumer prices had not been fully increased in line with international prices. As a result of the continuing expansion, the external current account and budget deficits reached 14.6 and 18.1 percent of GDP, respectively.

Under these circumstances, it was apparent that contractionary policy would be required for 1977. Accordingly, the Moroccan authorities doubled the special import tax from 4 to 8 percent, tightened credit, reduced some consumption subsidies, and slowed the rate of growth of public investment. As a result, the budget deficit was cut by over 3 percent of GDP, but the external current account deficit increased as nongovernment financial savings turned negative. By the end of the year, there was negative financial savings (that is, excess demand) in both the government and nongovernment sectors. Total external public debt had doubled in one year to US$4 billion. Debt service payments were projected to rise from 10 percent of exports of goods and services to 20 percent three years later. Average maturity was expected to decline sharply—by 50 percent to nine years in 1978 alone.

It was thus evident by the last quarter of 1977 that an even tighter policy stance would be required to keep the situation under control. Accordingly, the government prepared an austerity budget for 1978 as part of a financial stabilization program, supported in part by the IMF. The main issues were how to reduce the external current account and budget deficits in order to prevent an excessive buildup of foreign debt and financial crowding out by the Treasury.

1978-79 Stabilization Program

In putting together the stabilization program, the government had to judge whether it was preferable to achieve the desired improvement in the current account through export growth or import reduction. Export growth was expected to come mainly from phosphates and their derivatives, but only in the medium term since the underlying investments would take several years to complete. Furthermore, the prices of phosphates had recently plummeted. Nonphosphate export performance had been disappointing over the course of the 1973-77 Economic Plan, despite the creation of the export insurance and Central Bank export rediscount facilities. In short, the government considered that exports were unlikely to grow significantly in the near future. Prospects were judged to be little better in the case of remittances from Moroccan workers residing abroad. Consequently, it was felt that the current account deficit could be more effectively reduced through restricting imports. Such a policy would also enhance protection of new industries created in the wake of the 1973 investment codes. To prevent the development of protected monopolies, the government tried to avoid the use of import restrictions on capital goods, except for those that could be manufactured domestically.

On the budgetary front, it was planned to reduce the deficit through revenue enhancement and expenditure compression. As in 1975 and 1977, it was decided to raise the stamp duty and the SIT to 4 and 12 percent, respectively. On the expenditure side, the need to curtail investment was recognized except with regard to social, medical, and educational infrastructure. National defense expenditures were considered sacrosanct. Hence, a reduction of only 20 percent in capital expenditures could be planned. Little effort was made to contain recurrent expenditures. In particular, it was considered essential to increase budgetary expenditures on consumption subsidies to stabilize retail prices of certain basic commodities (flour, fertilizers, and petroleum). Continuation of "low" prices for these goods and certain services (public transport and household electricity) was considered politically essential.

These measures, however, proved quite insufficient to the task at hand. In the first few months of 1978, foreign exchange resources declined faster than forecast, and public expenditures exceeded available funds. At midyear, the government reacted by cutting capital expenditure by 50 percent rather than the initial 20 percent, issuing an even more restrictive annual import program (including a substantial increase in outright prohibitions), and introducing a six-month import deposit on about 30 percent of all imports. Importers received no interest for these deposits that were then onlent interest free to the Treasury. Finally, in order to spur worker remittances from France, a special exchange premium was introduced, so establishing parity between the dirham and the French franc, as against an official exchange rate of 1FF = 0.91DH. The implicit premium, paid by commercial banks and the Treasury, was rather larger than the similar premium of 5 percent paid on remittances originating in other countries.

In broad terms, this program was supported by the IMF, although the Fund differed with the Moroccan authorities on some points. In particular, the IMF contended that the budget deficit was attributable not only to excessive capital expenditures but also to the substantial growth of consumer subsidies and defense expenditures. Consequently, it stressed the need to raise domestic savings, both budgetary and private, through appropriate tax, expenditure and interest rate policies. Growth would be compromised so long as excessive consumption crowded out investment. Total budgetary expenditures must be contained within available financial resources including foreign and domestic borrowing (see Figure 2.9). Credit policy needed to be tightened without jeopardizing the legitimate needs of the private sector. On the external side, the IMF objected to the use of advance import deposits, arguing that it would have been better to raise import duties on a selective basis. Opposition to tighter import licensing was muted since this was expected to be transitory.

There were more substantial differences concerning interest rates, the exchange rate, and the application of premiums to worker remittances. The Fund argued that interest rates had become negative in real terms and that they should be raised to stimulate savings (both real and financial). But Moroccan authorities objected that this would only cause higher inflation and lower investment. The exchange rate, according to IMF estimates, had shown a slight tendency to appreciate, so hindering export development. The Moroccans retorted that the recent poor performance of exports should be attributed, not to a lack of competitiveness, but rather to nonprice factors such as poor climatic conditions and protectionism on the part of Morocco's trading partners. Furthermore, depreciation in the exchange rate would generate additional budgetary difficulties in the form of higher expenditures on debt service and consumption subsidies. Finally, the Fund considered the premiums on worker remittances to be equivalent to an undesirable multiple exchange rate practice, whereas Morocco argued that they were an effective stimulus to current transfers and enhanced national savings.

The overall macroeconomic results for 1978 were considerably better than for 1977 (see Figure 2.1a). But the intended momentum could not be maintained in 1979. Wage increases, higher expenditures on consumption subsidies, national defense, and other public investment led to a sharp increase in total public expenditure, as shown in Figure 2.8, and they generated the need for additional fiscal revenues. Hence, in July the SIT and the stamp tax were again raised, this time to 15 and 10 percent, respectively. The sales tax was augmented on certain categories of service transactions, and a temporary surcharge was introduced on wage and corporate income. In October 1979, the price of crude oil doubled, which put further pressure on the external current account and budget deficits. For the year as a whole, these were 9.6 and 10.1 percent of GDP, well above the amounts targeted in the 1978 adjustment program. External debt now amounted to US$5.2 billion, 20 percent higher than two years earlier. Debt service was expected to exceed a billion dollars for the next six years.

Under these circumstances, the improvements of 1978 and 1979 were considered to be transitory. Domestic savings had fallen to 10.6 percent of GDP. In the absence of further adjustment, the budget deficit was expected to increase to an unsustainable 13 percent of GDP. At the same time, domestic political pressures *against* adjustment were increasing in the face of consumer resentment, thus making it more difficult to reduce expenditures on consumer subsidies. Rising tensions in the western Sahara escalated military expenditures. These considerations, together with debt service projections, led the Moroccan government to conclude that the country's problems could not be solved by financial stabilization and demand management alone. Hence, it decided to request IMF support in the form of a three-year Extended Fund Facility. The purpose of the EFF, which incorporated both supply and demand measures, would be to help Morocco embark upon the growth-oriented structural adjustments needed to overcome worsening social and economic problems.

Stabilization and Adjustment, 1980-82

The IMF Extended Fund Facility (EFF)

The basic objective of the EFF was to considerably attenuate, if not eliminate, the external and budget deficits, which were considered unsustainable, by raising domestic savings. Investment also was expected to rise to enhance the growth rate. In this connection, the EFF aimed to provide SDR 810 million (about US$1 billion) over a three-year period. The loan was perceived as providing part of the US$2 billion of external resources required by the 1981-85 Economic Plan. The EFF's main objectives were to raise the domestic savings rate to 15 percent of GDP and to lower the budget and external current accounts to about 7.5 and 6.5 percent of GDP by 1981, from their 1979 levels of 10.1 and 9.6 percent, respectively. In addition, budgetary savings were to attain 30 percent of planned public investment, itself equal to a minimum of 6 percent of GDP. Annual and overall targets were specified for the balance of payments deficit and foreign borrowing.

These objectives were to be achieved through a combination of financial and economic measures to reduce demand and enhance supply. In line with IMF practice for standbys and extended facilities, the availability of each tranche was made conditional upon the satisfaction of periodically fixed and reviewed performance criteria. These were

- quarterly ceilings on total domestic credit of the banking system, with subceilings on net bank credit to the Treasury;

- a ceiling on new external loans from international markets, with maturities between one and ten years;
- an understanding that Morocco would not engage in multiple exchange rate practices or tighten payment restrictions over the course of the program.

The quarterly ceilings on total domestic credit were to be determined in the context of monetary policy (a) which limited the growth of money supply to the expected growth rate of GDP over the program period and (b) favored the nongovernment sector compared to the Treasury. On the occasion of each review, understandings would be reached concerning future exchange rate policy and suitable targets for the performance criteria.

The corresponding action plan was aimed primarily at containing public and private consumption. No increases in taxes were planned since additional fiscal revenues were expected to materialize from economic growth as well as from a comprehensive reform of direct and indirect taxes scheduled for implementation between 1981 and 1983. On the budgetary side, the objective was formulated in terms of limiting the rate of growth rather than the absolute level of expenditure, except concerning budgetary transfers to state enterprises and consumption subsidies that were to be reduced to their 1979 level in real terms (equivalent to a 5 percent reduction in nominal terms) by the end of the program. In other words, budgetary savings were to result mainly from higher prices of subsidized goods, which would restrain private demand. In the summer and fall of 1980, prices were increased between 10 and 50 percent. Further reductions in nongovernmental consumption were expected to result from larger savings, induced by the 20 percent increase in interest rates.

On the supply side, the main beneficial effects were expected to result first from a change in the composition of investment. Henceforth, 70 percent of all capital formation would be undertaken by the nongovernment sector, effectively reversing the 1973-80 trend. In addition, resources were to be channeled in priority to export-oriented sectors (citrus, vegetables, fisheries, and phosphates) and to economically efficient import substitution in industry and agriculture. Some attempt also was made to encourage more labor-intensive small-scale industries. Second, a major improvement in the financial performance of the public enterprise sector (600 enterprises) was envisaged, in order to reduce their financial dependence on the Treasury. The third component of the supply-side program concentrated on trade and exchange rate policy; the objective was to reverse the appreciation of the dirham since 1973. It was felt, however, that exchange rate depreciation could play only a limited role in Morocco because of the apparently low demand and supply elasticities. The EFF also aimed at avoiding any further tightening of import licensing and quantitative restrictions; the elimination of the advance import deposit; and the termination of the premium exchange rate paid on worker remittances. Finally, the EFF sought to increase the domestic supply of oil and gas, even though the benefits would not occur until after the program's expiration.

Performance Under the EFF

The EFF became effective in the last quarter of 1980. Despite some minor discrepancies between targets and outcomes, all performance criteria were met at the end of the year. In particular, expenditures on consumption subsidies had been kept within the targeted amount by increasing retail prices between 11 and 33 percent.

However, 1981 was to prove very difficult largely because of unanticipated events. In the wake of drought, agricultural output fell by 23 percent, which led to lower agricultural exports and higher imports of foodstuffs—particularly cereals, sugar, and edible oils. At the same time, the dollar prices of some of these commodities (for example, sugar) were rising, and the dollar itself was beginning to appreciate. As a result, budgetary expenditures on consumption subsidies rose sharply and were projected to reach 4.8 percent of GDP by year end (that is, more than two and a half times the amount targeted in the program and appropriated in the budget). The Moroccan government decided that the original targets should be respected. Consequently, it raised prices of petroleum in February (5 to 23 percent) and of other basic commodities (up to 70 percent) in May. But in the wake of violent consumer protest, the latter increases were approximately halved, and a compensating wage increase of 3 percent was granted. Despite another increase in petroleum prices in November, overall budgetary expenditure on consumption subsidies still exceeded the budgeted amount by 50 percent. The cost of debt service was rising on account of rising interest rates and the appreciating dollar. Plan-related capital expenditures overran their target by 24 percent. Consequently, the external current account and budget deficits reached 12.6 and 14.6 percent of GDP, compared with targets of 6.5 and 7.4 percent, respectively Foreign borrowing by the Treasury exceeded the agreed ceiling by 50 percent, thus violating one of the principal performance criteria. There was also a 3 percent overrun of the Treasury's domestic credit ceiling.

In the light of these factors, it became clear that the EFF's targets could not be achieved within the original time frame. The program was therefore canceled at the end of 1981 and replaced by an eighteen-month standby and a drawing under the CFF facility. However, there was no loss of resources to Morocco, and it was thought that another EFF might be negotiated at the end of the standby if this went well.

IMF 1982 Standby

Corresponding to its perceived role as a bridge between the failed 1980 EFF and a new one that might begin in 1983, the 1982 standby aimed at combining short-term stabilization of the budget and external current account deficits with the continuation of the EFF's structural reforms on public enterprises, taxes, education, etc. The performance criteria were the same as those of EFF, namely:

- quarterly ceilings on total bank credit, with a subceiling for bank credit to the Treasury;
- a ceiling on new nonconcessional external loans contracted or guaranteed by the Moroccan government in the maturity range of one to ten years;
- a midterm review of exchange rate and interest rate policies;
- the standard IMF provisions relating to multiple currency practices.

In addition, quarterly targets (not legally binding) were introduced for government revenues and expenditures to improve identification and monitoring of the program.

The targets were supposed to be achieved through a combination of revenue-raising and expenditure-reducing measures. On the revenue side, the standard rate of the manufacturer's sales tax (TPS) was raised from 15 to 17 percent, and excises on selected commodities were adjusted upward. From these and other ancillary measures, additional revenues of about 1 percent of GDP were expected to materialize. Total expenditure was expected to increase by only 1 percent, equivalent to a 12 percent reduction in real terms. Public sector wages were frozen. Interest rates were raised in April 1982, and the exchange rate was further depreciated. Structural reforms were to be maintained. It was again promised that the enabling legislation (loi cadre) for the tax reform would be submitted to Parliament, and that the net transfers to public enterprises would be reduced. In that connection, prices of some public utilities were increased (irrigation water by 16 percent, drinking water by 15 to 20 percent, rail transport by 10 to 15 percent, and electricity by 28 percent).

Progress in the early months of the standby was satisfactory, but by August it was clear that the year-end targets could not be met primarily because of a shortfall in phosphate exports (26 percent) relative to their projected value. As a result, the budget deficit was projected to attain 9.3 percent of GDP rather than 8.2 percent, and the external current account 12 percent of GDP instead of 10 percent. The actual overruns were even greater, with the budget deficit reaching 12.6 percent of GDP, almost 54 percent over target. The external current account deficit was 12.7 percent of GDP, about 25 percent more than planned. The overrun for the overall balance of payments was 43 percent. These excesses notwithstanding, all performance criteria were met essentially because the government obtained additional *concessional* foreign borrowing, for which there was no ceiling in the program. But external arrears were accumulated, and foreign exchange reserves were run down to just over two weeks of imports.

At the end of 1982, external debt had grown to about 78 percent of GDP. Debt service was projected to exceed 40 percent of exports of goods and services including worker remittances. Nevertheless, the 1983 budget was highly expansionary and implied a deficit of 20 percent of GDP, even greater than in the euphoric years of 1976-77. In 1983, however, new foreign borrowing was not as available as in the earlier period. A financial crisis emerged in March when foreign exchange ran out. Import licenses were reintroduced on all goods, together with advance deposits in some cases. Emergency financial assistance, including debt rescheduling, was requested from bilateral and multilateral sources.

World Bank Structural Adjustment Loan

In 1980-81, the World Bank attempted to put together a structural adjustment loan (SAL) of about $500 million. This operation, the Bank's first foray into policy lending in Morocco, was thought of as a supply-side operation that was intended to complement the EFF and to provide part of the external financing required by the 1981-85 Economic Plan.

The analytical basis of the SAL was the *Basic Economic Report* (World Bank 1979), which found that the Moroccan economy had made considerable strides, especially since 1973. However, maintaining economic growth and improving living standards for the poor required larger investment expenditures that would have to be financed out of higher domestic savings and exports. Therefore, the report argued that there was a need for stabilization to raise domestic savings, and especially budgetary savings, as well as a need for adjustment to enhance the economy's capacity to earn foreign exchange, especially from exports. Export growth had recently slowed from 8 to 1 percent, in part because of the increasing anti-export bias that resulted from the import substitution policies of the 1973-77 Economic Plan.

Consequently, the *Basic Economic Report* argued that the top priority for economic policy was the preparation of the structural reforms needed to enhance economically efficient growth in productive sectors and exports. In particular, investment incentives needed to be reoriented toward production for export. Import substituting industries needed to operate more efficiently and at lower cost through continuous (downward) adjustment of protection. In this connection, considerable emphasis was placed on the need to better understand price policy. Rigid controls in some sectors coexisted with complete freedom in others. The problem was particularly acute in the public enterprise sector where price controls often generated operating losses that had to be financed by transfers from the Treasury. Reduction of these losses was a key component of the program to raise budgetary and hence domestic savings.

The SAL attempted to build an adjustment program on the basis of the preceding developments. This program was intended to improve the sectoral allocation and efficiency of investment, enhance private and public savings, and increase exports. Yet throughout preparation of the operation, it proved difficult—even within the Bank—to agree on the specifics of the program. In the final phase of negotiations, the following conditions were specified:

- a minimum 10 percent economic rate of return for public sector investments;
- elimination of operating and capital transfers from the budget to public enterprises (in addition, public enterprises should finance one-third of their investments from internally generated funds);
- an increase in interest rates except for agriculture;
- a reduction in fertilizer and feed subsidies by 50 percent before 1983.

Other important objectives included raising the domestic savings rate from 11 to 16 percent. Since investment was expected to average about 25 percent of GDP, foreign borrowing would make up the difference—whence the financial justification of the SAL.

In preparing this loan, the Bank worked closely with the IMF, but the Fund's quantitative targets were intentionally not incorporated into the SAL as performance criteria. As a result, the SAL contained no specific macroeconomic conditionalities although the Bank impressed upon the IMF the importance of maintaining some of Morocco's ongoing social programs. Rather, the SAL focused on investment patterns, export promotion, and sectoral policies. Excluded from the SAL were objectives that could be better met through traditional projects. Also, there were no conditionalities relating to price policy and regulation—a striking omission given the attention paid to price policy in the *Basic Economic Report* and the impossibility of reducing budgetary transfers to public enterprises without increasing prices.

In the end, agreement could not be reached with the Moroccan authorities concerning any of the preceding conditionalities. Consequently, the negotiations were suspended and the SAL abandoned.

A Comparison of Approaches to Adjustment, 1973-82

The Moroccan Government's Approach

The government initially attempted to mobilize domestic resources to finance the deficits as they escalated in 1976-77. It tried to contain domestic demand through restrictive monetary policy and across-the-board increases in import levies (namely, the special import tax and the stamp duty). But by 1978, stronger medicine was prescribed in the form of higher import levies, cuts in public investment, and tighter import restrictions. These policies were reinforced in 1979. The government, however, failed to evaluate their impact on production incentives. In particular, their increasing anti-export bias was completely overlooked.

By the end of 1979, the government recognized that Morocco's problems could not be solved by demand restraint alone, even though this was indeed essential. Supply-side measures were needed as well. The measures adopted in the 1981-85 Economic Plan expanded investment, often with long gestation periods and high capital output ratios. Little attention was given to improving producer price policy and/or the regulatory framework as it related to production. Reducing anti-export bias and export promotion were also ignored.

As a result, the government's program was essentially oriented toward demand restraint even after 1980. However, the revenue-raising measures began to assume the economically more desirable form of higher internal taxes rather than protective import levies. Thus, increasing anti-export bias was at least avoided. But the political pressures to expand social programs for investment were too great. As a result, the momentum of stabilization and adjustment could not be maintained. This led to the expansion of the 1983 budget and the financial crisis of March 1983.

The IMF's Approach

From 1975 to 1979, the IMF seems to have been in basic agreement with the Moroccan government's attempt to stabilize the situation by reducing the budget deficit through increased taxes and reduced expenditures. Not too much attention was given to the microeconomic consequences of the increases in the SIT and the tighter import restrictions.

Like the government, the IMF realized by the end of the 1970s that demand restraint was not enough to solve Morocco's problems and that a supply-side program also was needed. To this end, the Fund and the authorities negotiated an Extended Fund Facility program, based on the IMF's financial programming model that sets ceilings for key variables (namely, total domestic credit of the banking system, bank credit to the Treasury, and nonconcessional foreign borrowing with maturities between one and ten years). In addition, it was agreed that annual understandings would be reached about the conduct of exchange rate policy, and that a tax reform would be adopted to increase government receipts, partly through a broadening of the tax base.

With the possible exception of exchange rate policy, there were no specific supply-oriented measures or performance criteria. Moreover, on account of elasticity pessimism, the anticipated benefits from the new flexible exchange rate policy were not considerable. The emphasis on investment suggested that the supply-side effects would be felt only in the medium term, long after the EFF's conclusion. Thus, the bulk of the performance criteria and program measures related, in fact, to demand management.

The success of the program depended on the control of public and private expenditure through reduction of consumption subsidies. Such a reduction was necessary to avoid violating the ceilings on domestic credit and foreign borrowing. However, the extent of the cuts in public expenditure needed to attain the targets resulted in substantial social unrest in May 1981. The price increases were partially rescinded. In turn, this caused the EFF program ceilings to be violated and led to the program's eventual cancellation in late 1981.

Nevertheless, in the early part of 1982 a new standby, also based on the Fund's financial programming model, was concluded with conditionalities similar to those of the EFF. This program was kept on track essentially because the Moroccan government managed to meet its foreign borrowing requirements through concessional loans. The program's targets were stipulated in terms of nonconcessional borrowing. Thus, the excess volume of foreign borrowing was not a technical violation of the program's conditionalities. Moreover, it was not recognized that the Treasury deficit targets were being met through an accumulation of arrears to the domestic suppliers, both public and private. This, in turn, effectively froze part of the private sector's use of bank credit, reducing the amount available within a given credit ceiling for production. Clearly, some refinements in the Fund's approach to stabilization and adjustment were needed. They were incorporated into post-1983 operations (see Chapter 5).

The World Bank's Approach

The attempted structural adjustment loan revealed perhaps the most comprehensive view of Morocco's problems. The Bank perceived that the budgetary and current account deficits of the late 1970s masked much deeper structural problems that required not only major investments but also reforms of economic policy, in particular relating to prices. Yet, in retrospect, the SAL's conditionalities were not on a par with its analysis. In fact, while somewhat broader than those of a traditional investment project, its conditions were noteworthy for their diffusion, their generality, and their relation primarily to public investment. Although the *Basic Economic Report* emphasized the probable inadequacy of existing price policy, the SAL eschewed any specific conditionalities related thereto. This was all the more surprising given the view of the Bank's Loan Committee that price policy in the agriculture sector could constitute a major issue in structural adjustment loans.

The SAL was not based on a sufficiently well-defined central theme to enable the definition of precise conditionalities pertaining to structural reforms. This is important because the next round of structural adjustment operations in the industrial, agricultural, and public enterprise sectors recognized that a fundamental problem in Moroccan economic policy was the structure of prices and production incentives both for exports and import substitutes. As the concluding chapter will explain, attention would finally turn away from investment to issues related to production, specifically price incentives and financing.

5
Adjustment, 1983-87

This chapter begins with a discussion of three IMF standby loans to Morocco between September 1983 and February 1988. The chapter then examines sectoral adjustment operations of the World Bank in industry, agriculture, and the public sector. The principal emphasis is on the preparation and performance of the Bank's Industry and Trade Policy Adjustment (ITPA) programs.

IMF OPERATIONS

The 1983-85 Standby

As indicated in the preceding chapter, the IMF's 1982-83 standby was fully disbursed despite the deterioration of Morocco's financial situation in the last quarter of 1982 and the external payments crisis of March 1983 brought on, in large part, by an overexpansionary budgetary policy. Initially, the Moroccan government believed that the crisis could again be ridden out by recourse to concessional external assistance.

It rapidly became clear, however, that external financing would not be as available as in the past. Consequently, there would not be enough resources over the next few years to make full debt service payments and to maintain imports of goods at desired levels. Debt rescheduling was perceived to be indispensable yet hardly obtainable without IMF support. The IMF had lent Morocco almost US$1 billion since 1980. Repayments would soon start to fall due. Maintaining or even increasing exposure would be difficult without rescheduling and perhaps even new inflows from bilateral and commercial debtors. These considerations gave the Moroccan government and the IMF a strong incentive to conclude a new standby. Considered the first of several, the standby was to be cast in a medium-term framework.

The standby, amounting to SDR 300 million, covered the period from September 1983 to February 1985. It was designed to mesh with the World Bank's Industry and Trade Policy Adjustment (ITPA) loan that was expected to become effective late in 1983 or early in 1984. The standby was considered part of a longer term program designed to reduce the economy's resource gap from 15.3 percent to 3 percent of GDP by 1987. The achievement of the longer term objective was an important factor in designing the standby's annual targets and actions.

There were some notable new performance criteria. These included quarterly ceilings on net borrowing by the Treasury (including variations in the outstanding stock of arrears) and domestic bank credit (with a subceiling for net bank credit to the Treasury). Annual ceilings were imposed on external debt contracted, guaranteed, or authorized by the government.

An innovative performance criterion was to call for periodic reductions in outstanding Treasury arrears. This implied a target for the Treasury's cash deficit after arrears repayment. The external debt performance criterion was stricter. It stipulated a ceiling for loans with maturities in the one-year to fifteen-year range compared with only one to ten years in earlier programs. For the first time, a ceiling was included for short-term debt having maturities of fewer than twelve months.

Targets and Outcomes

In 1983, the main objectives were to attain (1) an external current account deficit of 9 percent of GDP and (2) an overall budget deficit of 9.2 percent of GDP before rescheduling and arrears repayment. In addition, the latter were to be reduced by about 2 percent of GDP by year end. These objectives were to be attained via a combination of actions in the area of budgetary, monetary, trade, and exchange rate policy.

Regarding budgetary policy, the manufacturer's sales tax (TPS) was increased in July from 17 to 19 percent. Capital expenditures were slashed. Explicit and implicit consumption subsidies were reduced by forcing through politically difficult price increases of 18 to 70 percent in many basic consumer goods and services, including water, electricity, and petroleum products. A comprehensive tax reform was to be submitted to the October session of Parliament, with implementation of the value added tax (VAT) scheduled for mid-1984.

As regards the trade and exchange regime, the primary objective was to maintain the flexible exchange rate policy instituted in 1980. This was to be accompanied by a whole gambit of export promotion and import liberalization measures, also recommended by the World Bank as part of its forthcoming ITPA loan. These included, inter alia, improving customs procedures on the export side, easing import licensing requirements back to their end-1982 level, eliminating advance import deposits, and lowering the maximum rate of import duties.

Monetary and financial policies were also revised, in the context of generally tight credit expansion targets; higher interest rates including a special rate (8 percent for worker remittances); and a broadening of the money market to improve the allocation of financial and Central Bank resources.

All performance criteria were satisfied at the end of 1983, except for a negligible overrun in bank credit to the Treasury. In addition, as required by the IMF as an implicit condition of implementing the standby, significant rescheduling was obtained for commercial and bilateral donors. Moreover, a pool of bilateral donors committed themselves to the provision of exceptional balance of payments assistance over the coming years.

The same restrictive policy stance was planned for 1984, in order to consolidate the gains made the year before, except that no revenue-raising measures were foreseen. But 1984, like 1981, proved difficult and disappointing.

The standby had first to take account of ITPA's requirement of a 5 percentage point reduction in the special import tax (SIT) on January 1, as well as a cut in the maximum rate of duty from 180 to 60 percent on July 1. The corresponding revenue shortfall was estimated at 1.5 percent of GDP. This was reflected in the standby's performance criterion that allowed an increase in the Treasury deficit equal to the expected value of ITPA's disbursements.[1]

However, the original targets for the current account and the budget deficit could not be met for a number of reasons. Exports showed lower growth and imports higher growth than expected. The additional imports (13 percent compared with a target of 2 percent) emanated from the recurrence of drought, which increased demand for wheat and petroleum, as well as from the massive investments of the OCP. The budget was also adversely affected by renewed social unrest following another round of consumer price increases. However, as in 1981, some of these increases—especially of low-quality flour and sugar that had already been spared in August 1983—were rescinded or postponed. Consequently, outlays on consumption subsidies were projected to rise to 2.5 percent of GDP; and, accordingly, various internal taxes and revenues were raised by about 0.75 percent of GDP. However, further upward pressure in consumer subsidy outlays resulted from the appreciation of the US dollar. This was supposedly palliated by a cut in capital expenditures of about 0.5 percent of GDP.

As a result of these unforeseen events, the target outcomes for the budget deficit (after rescheduling) and for the external current account (before rescheduling) were revised upward. The increased current account deficit implied an increase in foreign borrowing, which remained subject to an unchanged ceiling. The government skillfully avoided the risk of noncompliance with program conditionalities by persuading certain of its bilateral creditors to grant offsetting increases in debt relief and concessional borrowing, which were exempt from ceilings under the standby. As a result of these maneuvers, the standby's original performance criteria were all effectively satisfied.

Comments

Yet the outcome for the year as a whole could hardly be considered satisfactory. The external deficit had increased rather than decreased. Official foreign exchange reserves, still barely equal to one week's imports, had not been reconstituted as planned. Budgetary expenditures on consumer subsidies had increased in the wake of the failure to raise consumer prices as much as desired. The reduction in capital expenditures proved more imaginary than real. Only cash payments for this class of expenditure were reduced, while the underlying transactions were maintained and financed by accumulating arrears. Furthermore, it emerged that the total of corresponding arrears for preceding years might be as high as 22 billion dirham—that is, over 20 percent of GDP. Finally, exchange rate policy was less than totally satisfactory as depreciation in the semester was reversed in the second half of the year.

---

1. As will be explained later, the World Bank expected that the revenue shortfall would be partially compensated in the short run by a series of increases in the TPS, as had occurred in mid-1983, as well as by increased transfers from the national phosphate company, Office Cherifien des Phosphates (OCP), and in the medium term by the additional revenues resulting from higher growth of exports and GDP.

Two More IMF Standbys, 1985-87

Targets and Outcomes

Following completion of the 1983-85 standby in February 1985, the government requested further assistance from the IMF consisting of a standby (SDR 225 million) and a drawing under the Compensatory Financing Facility (SDR 85 million). The latter was to compensate for shortfalls in phosphate exports and cereals production.

The design of the program and its performance criteria reflected an increased understanding of the complexity of Morocco's economic problems, in particular the need to restore sustainable internal and external balances no later than 1988. Accordingly, target ratios, relative to GDP, for the current account and Treasury deficits were defined as follows for 1985 and 1986:

	1985	1986
External current account before debt relief	6.5%	4.5%
Budget deficit		
Payments order basis, before debt relief	6.5%	n.a.
Cash basis, after arrears reduction	8.9%	4.5%

In 1985, arrears repayments were to attain 2 percent of GDP. Foreign exchange reserves were to increase to two weeks of imports by the end of the program.

Reflecting the greater appreciation of the problems at hand, the program's performance became correspondingly more complicated. In addition to the by now conventional ceilings on total domestic credit and bank credit to the Treasury, legally binding targets were introduced for allowable external arrears, as well as outstanding bills awaiting payment. The definition of the budget deficit target became more sophisticated and included a quarterly ceiling on a cash basis before debt relief excluding reduction in domestic bills awaiting payment.

Combined with a declining ceiling on the allowable stock of such bills, this effectively implies a quarterly target for their net reduction. In turn, this ties into the relevant target ratio for the overall Treasury deficit, relative to GDP, on a cash basis after arrears reduction, to be attained by the end of the program.

As in 1984, policy actions under the proposed standby had to be coordinated with the exigencies of the ITPA operations, particularly regarding fiscal and trade policy. Of particular significance in this regard was the reduction of the SIT from 10 to 7.5 percent on January 1, 1985, which entailed a revenue loss of about 800 million dirham (that is, about 0.67 percent of GDP, or 3 percent of total revenues).

Surprisingly, no compensating revenue measures were envisaged in the 1985 budget laws. This was perhaps because revenues were already forecast to rise by about 16 percent (800 million dirham were expected to originate from higher budgetary transfers by OCP). A major effort was expected on the expenditure side, with strict limits on the increases in government wages and consumer subsidies. Capital expenditures were to be contained to levels that avoided a buildup of arrears.

Monetary policy was designed to accommodate the budgetary developments, while providing a modest increase in credit to the private sector. Interest rates were raised across the board by 1 to 2 percentage points. Further reforms in financial markets were initiated, in particular the encouragement of financial savings via the elimination of all obligatory placement requirements on long-term deposits.[2]

External policies (flexible exchange rates, import liberalization, and export promotion) were to be continued in order to buttress the ITPA programs. In addition, external arrears were scheduled for elimination by the end of July. Rescheduling of all 1985 bilateral and commercial debt service obligations was requested.

However, developments in the months immediately following the standby request were broadly unfavorable. At the end of June, budgetary expenditures, especially on capital, considerably exceeded targets. External arrears increased because of delays in disbursements of new foreign loans. Imports of cereals and petroleum increased, despite an excellent harvest. Export performance was also disappointing because the 50 percent decline in phosphoric acid earnings was only partially offset by the 15 percent growth in other exports.

Consequently, a revised program was prepared, based on a slightly larger external current account target and a correspondingly smaller budget deficit—resulting from enhanced revenue collection and further increases in consumer prices. The revised program was finally negotiated in September.[3] But within three months there were major breaches of the performance criteria and understandings, especially regarding the budget. The overall budget deficit (on a payments order basis, before debt relief) was 50 percent above target, and arrears repayment was significantly below. The external current account deficit was also 9 percent larger than expected, and Bank Al Maghrib's reserves were increased only by half the targeted amount. Bank credit ceilings for the Treasury and the economy were also largely exceeded. The standby was therefore canceled after a small initial drawing, although drawings under the CFF continued uninterrupted.

These developments caused considerable alarm in some official circles. It was argued that the adjustment programs—particularly regarding the exchange rate, import liberalization, and the reduction of import levies—were causing Morocco's current account and budget deficits to increase rather than go down. Strong pressures began to build for a slowdown in the pace of adjustment.

Yet developments in the external sector were arguably positive despite noncompliance with program targets. Nonphosphate manufactured exports increased by 15 percent in volume terms compared with only 9 percent the year before. Tourism earnings recorded growth of 30 percent, and worker remittances advanced considerably. However, overall export growth was set back by an 8 percent decrease in phosphate earnings owing to depressed phosphoric acid prices. At the same time, total imports declined by 1.5 percent in real terms. However, import performance was, in fact, better since total imports included foreign raw materials that were reexported after processing. These increased significantly

---

2. Commercial banks were required to place about 40 percent of their deposits in Treasury bills and obligations of specialized institutions, both of which paid less than market rates of interest.

3. The revised standby was reduced to SDR 200 million, but the CFF increased to SDR 115 million. Hence, the amount of resources made available to Morocco actually increased by SDR 5 million.

both volume and value terms.[4] Hence, imports for domestic consumption fell even more than total imports.

The Ministry of Finance contended that the poor budgetary developments were principally attributable to the reduction in the SIT and, in lesser degree, to a certain lack of buoyancy in other categories of tax receipts. It argued strongly that any further reduction in import duties and the special import tax should be eschewed unless the revenue shortfall could be made up by an offsetting increase in domestic tax receipts. But these could be expected to emerge only after implementation of the long-awaited tax reform, then scheduled for April 1986, which would notably reduce exemptions from direct taxation of profit-related income and from indirect taxes. Furthermore, it was held that expenditures could not be cut further. Increasing prices continued to pose insurmountable sociopolitical problems. In addition, it was proving increasingly difficult to contain commitments on capital expenditures to levels consistent with available financial resources. Under these circumstances, further cuts in taxes would cause further buildup in arrears, both domestic and foreign.

Similarly, in discussing developments in the external sector, the Ministry of Finance argued that the increase in consumer goods imports was entirely attributable to the liberalization of competing imports. It did not note that some of these imports (for example, fabrics) were transformed into manufactured exports as were some intermediate goods (for example, yarns, skins, hides, sulfur, ammonia and cotton).

Despite the cancellation of the 1985 standby, the government continued to act as if one were in force. Indeed, the intent was to obtain IMF assistance again as quickly as possible. To this end, a target deficit of 6.6 percent of GDP on a cash basis was set and attained for 1986. Principal measures were to maintain the SIT at 7.5 percent (instead of reducing it to 5 percent as required by the World Bank) and to institute an exceptional fiscal levy to capture the windfall gain from the January decline in the international price of petroleum.

A new fifteen-month standby of SDR 230 million was finally approved in December 1986. The program design was roughly the same as the preceding operation. The main objectives were approximate equilibrium in the current account, the elimination of all external arrears, and an increase in official reserves to about one month of imports. The budgetary target was a further reduction in the overall deficit on a cash basis before relief to 5.3 percent of GDP. The medium-term objective was formulated as reducing the debt service ratio, before debt relief, from 70 percent of exports of goods and services to a more reasonable 30 percent by 1993.

As indicated, the program's design was roughly the same as in the preceding standby. The most notable measure was the decision to formally oppose any further reduction in the effective incidence of import duties and their equivalents. Consequently, a reduction of 2.5 percentage points in the SIT undertaken in January 1987 was offset by an equivalent increase in the customs duty. This compensatory measure directly violated a covenant of the ITPA loans, which proscribed raising any other non-neutral taxes to offset the reduction of the SIT. In addition, a variety of other revenue-raising measures were taken with a yield of

---

4. The volume effects resulted from the overall growth of exports in general. A particular influence was massive purchases of sulfur by OCP undertaken in part because of the surge in its world market price. This is the price effect referred to in the text. Also many of these imports were textile fabrics that are counted as consumer goods.

about 400 million dirham.[5] Reductions in recurrent expenditures were also proposed in order to avoid any further reduction in capital expenditures.

Monetary and exchange rate policies were designed to buttress these macroeconomic objectives, through restrictive credit policy and the pursuance of a flexible exchange rate policy. Similarly, continuation of import liberalization was planned.

Comments

From this analysis it appears that a substantial increase occurred in the scope of IMF programs. Performance criteria became considerably broader and more detailed than in the earlier period, essentially on account of the need to address the enormous internal arrears problem, which had been neglected before 1983. Of course, settlement of the arrears problem called for even more draconian budgetary stabilization. In turn, this implied a correspondingly greater need for budgetary resources and made any reduction in import levies that much more difficult to support. At the same time, the program also introduced tragets for official external foreign exchange reserves. Unfortunately, this condition was not satisfied despite the desirability of adequate external reserves for macroeconomic management.

In terms of the methodological approach presented in Chapter 3, the post-1983 IMF standbys continued to emphasize budget deficit reduction as the primary means of bringing the external current account back to more manageable proportions. Compared with the pre-1983 era, more attention was paid to exchange rate depreciation as a way of stimulating export growth. The standby programs remained, however, essentially macroeconomic in nature with relatively little effort devoted to detailed microeconomic questions.

WORLD BANK OPERATIONS

Beginning in 1983, the World Bank became a major financial supporter of structural adjustment in the industrial, financial, public enterprise, agricultural, and education sectors. The following discussion concentrates on the two loans (ITPA1 and 2) that financed industry and trade policy adjustment. It first focuses on the economic analysis underlying these operations, which showed that Morocco's macroeconomic problems of the late 1970s and early 1980s reflected not only excess demand but also major distortions in the structure of production, trade, and fiscal incentives. These distortions were causing inefficient allocation of resources, thereby aggravating the macroeconomic problems. At the same time, there was an inadequate supply of foreign exchange and of domestic and financial savings relative to the investment and production needs of industrial sector firms, particularly exporters and their domestic suppliers. Following this analysis, the preparation and negotiation of the two loans are discussed. The section concludes with an evaluation in macro- and microeconomic terms that draws on the methodology developed in Chapter 3.

---

5. On political and legal grounds, the government again rejected increases in the National Solidarity Levy (PSN).

Analysis

Macroeconomic and Exchange Rate Developments, 1973-82

As Table 2.3 showed, the current account and budget deficits worsened considerably between 1973 and 1982, when they reached the unsustainable levels of 12.7 and 12.6 percent of GDP respectively. Domestic savings had declined from 15.6 to 8.4 percent of GDP, as a result of declining trends in both government and nongovernment savings. Indeed, government savings had become negative by 1981.

At the same time, the real exchange rate had appreciated between 20 and 25 percent in the 1970s—hardly desirable since, during the same period, the terms of trade had shown a net depreciation with the exception of the 1974-76 boom (see Figure 2.6 and Annex 2, Table 9). Moreover, export performance had been rather poor, and Morocco had lost market share to its competitors in both goods and services, especially tourism.

In other words, it appeared that the domestic supply of foreign exchange was not growing in line with the economy's needs and might indeed be stagnating.

Price and Production Incentives for Exports and the Domestic Market

One of the most important features of ITPA was that it was based on a careful quantitative analysis of the relative structure of price and nonprice incentives for exports and the domestic market. On the export side, the main disincentives were explicit or implicit commodity-based levies. Most paid a statistical export tax of 0.5 percent. Processed foods paid a somewhat higher rate because they had to be exported under the auspices of the Office de Commercialisation et d'Exportation (OCE). The OCE, now virtually abolished, functioned like a state marketing board, whose total levies were about 5 percent of export value—considerably in excess of the cost of the services rendered. In addition, many commodities were subject to de facto export licenses delivered by the Exchange Control Office. In general, export licenses were used to prevent exports occurring at less than officially mandated prices, usually determined on an ad hoc basis.

The primary focus of concern regarding incentives for import substituting industries was the persistently rising trend in tariff and nontariff protection between 1975 and 1979. Four contributory factors help explain this trend:

- increases in import duties, generally following requests for protection;
- increases, generally for budgetary reasons, of two other import levies, namely the special import tax (from 2 to 15 percent) and the stamp duty (from 1 to 10 percent);
- the increasingly restrictive nature of the import licensing regime;
- the introduction of advance deposits on imports.

These factors had led to increased anti-export bias, characterized by unit value added (or transformation) margins that were lower for exports than for sales to the domestic market. This anti-export bias was reinforced by the overvaluation of the exchange rate, relative to its equilibrium value, resulting from the use of tariffs and import licenses to contain import demand.

All of the above had been carefully documented in a firm-level study of the structure of incentives, prices, and economic efficiency in the industrial sector. This study, which used the effective protection study methodology described in Chapter 3, was undertaken jointly by the World Bank and the Ministry of Commerce and Industry between 1979 and 1982. It covered a sample of about 200 firms, with quantitative indicators of protection being estimated for about 85 firms in 1978 and later when possible.

Three indicators were calculated: the nominal protection coefficient (NPC), the effective protection coefficient (EPC), and the effective subsidy coefficient (ESC). These coefficients were first estimated relative to the official exchange rate and then, in net terms, adjusted to take account of any overvaluation. The study measured the efficiency of resource allocation using the domestic resource cost (DRC) of foreign exchange.

Other important aspects of price and incentive regimes were highlighted by the study. The first concerned domestic regulation (price control, domestic content requirements, barriers to entry and exit, labor market regulations); availability and cost of investment finance and working capital; the investment code (restrictions against foreign investment, interest rate subsidies and other sources of bias against employment, desirability of tax holidays, complexity of approval procedures); and the structure of indirect and direct taxes (distortions and levels of taxation particularly regarding choice of technique and the cost of capital and labor). The second major theme related to the inefficiencies of export-related special customs as well as indirect tax regimes and, more generally, all international trade procedures (port clearance procedures, customs forms, bank practices, telecommunications, as well as internal and external transport). The purpose was to identify the main nonprice administrative and institutional barriers to export development. These turned out to be of considerable importance and have been followed up in subsequent World Bank operations, where the issue of export marketing has also received major attention.

INDICATORS OF EFFECTIVE PROTECTION BY MARKET. Firms' unit transformation or value added margins, measured in domestic prices, averaged about 50 percent higher for domestic sales than for exports, according to the study. These margins for exports were less than for border prices; the reverse was usually the case on the domestic market. In other words, relative to border prices, production for export was implicitly taxed. For the domestic market, production generally benefited from

Table 5.1 Effective Protection Coefficients by Market

Sector	Domestic market	Exports
Textile and leather	1.21	0.88
Mechanical and electrical	1.35	0.87
Chemical	1.22	0.54
Agro-industry	1.12	0.88
Average	1.25	0.83

Source: Government of Morocco (1983a).

protection-generated subsidies to unit value added margins, although not for some intermediate and capital goods. Table 5.1 summarizes the results of the effective protection analysis.

These figures confirm the relative taxation of production for export compared with domestic sales. Although not particularly high, the EPCs for the domestic market are production-weighted averages and as such mask considerable dispersion.[6] In some instances, the EPCs were very high, but these were offset by the lower EPCs observed for other industries, which had a high weight in the sample. Some of the "high" EPCs for the domestic market are shown in Table 5.2.

Table 5.2 Domestic Market EPCs for Selected Industries

Industry	EPC
Foundries	1.86-2.64
Reinforcing steel rods	1.93
Galvanized pipes	3.94
Galvanized wires	32.10
Kitchen stoves	3.53
Refrigerators	10.25
Automobiles	1.58
Televisions	3.79
Cables	2.01
Diesel motors (stationary)	1.82
Paper pulp	1.88
Paper	2.08
Ceramic tiles	2.66
Glassware	1.71
Caustic soda, chlorine, and PVC	5.96
Yeast	2.99

Source: Same as Table 5.1.

However, EPCs do not give a complete picture of the structure of incentives since they only take account of tariff and nontariff barriers to imports. But in the Moroccan case, firms received tax or financial incentives through the investment and export codes or were penalized because of restrictions, in the case of agro-industrial exports, on their choice of marketing agent. The effective subsidy coefficient adjusts the EPC to take account of these factors.[7]

---

6. On occasion, EPCs in the sample were negative, indicating that the activity or product in question entailed a loss of foreign exchange for Morocco. That result also occurred in some of the more detailed analyses of petitions for protection, which were undertaken by the incentives unit at the Minister of Industry's request.

7. See Chapter 2 for a description of the main investment code benefits. The main export-related incentives and disincentives were abatement of profits tax, prorated to the share of

On the domestic market, the ESC was systematically greater than the corresponding EPC except for cement factories, which paid heavy levies to the Price Stabilization Fund. This relation between the ESC and the EPC indicates that the investment code augmented incentives for firms to produce for the domestic market.

For the export market, the results of the ESC analysis were more complicated and may be classified as follows:

Group A: Garment making   ESC > 1 > EPC

Group B: Agro-industries   ESC < EPC < 1

Group C: Other industries   1 > ESC > EPC

Thus, the different export promotion measures more than offset the implicit taxation of exporting as represented by an EPC of less than unity only for garment making. These firms received a net subsidy in effective terms, as indicated by ESCs exceeding unity. For the second category of industries (Group B), the ESCs were less than the EPCs; in other words, the effective taxation of exporting was greater than indicated by the EPC despite the export code's profits tax abatement and the interest rate subsidies given by the Bank of Morocco. All agro-industrial firms paid the OCE a commission equal to 3 percent of each export sale, and this commission was effectively a tax rather than a fee for services rendered. Finally, for the third group of firms, the ESCs exceeded the EPCs but remained less than unity. This indicates that the export promotion measures attenuated but did not eliminate the implicit taxation of exports resulting from the structure of protection as measured by the EPCs.

Both the EPCs and the ESCs reported above have been calculated at the official exchange rate. This, however, was clearly overvalued because imports were being effectively restricted by tariff and nontariff barriers. As a result, the preceding incentive indicators underestimate the true degree of effective taxation of exports and overestimate the effective subsidization of the domestic market. The error involved is approximately equal to the degree of overvaluation that was calculated to be about 15 percent. Dividing the preceding EPCs and ESCs by 1.15 gives the net EPCs and ESCs (see Table 5.3).

The main inference is that, on average, production for the domestic market was subsidized in net terms—even after adjusting for the overvaluation of the exchange rate.

The overall conclusion of the EPC-ESC analysis was that production for all export activities was unambiguously taxed in effective terms. On the domestic market, however, some activities (capital and intermediate goods) were also effectively taxed just like production for export, whereas others were subsidized in that EPCs or ESCs or both were greater than unity. Yet the production-weighted average EPCs and ESCs were greater than unity, indicating that domestic market sales were, on average, subsidized. These conclusions held even after correcting for the estimated overvaluation of the official parity of the dirham.

---

exports in sales; low-cost Central Bank pre- and postshipment working capital credits; limited duty-free imports of raw materials; a duty drawback scheme; a small export insurance facility; and an obligation on agro-industrial producers to channel their exports through OCE. The last two acted, in some cases, as disincentives.

ECONOMIC PROFITABILITY OF ACTIVITIES. The World Bank/Government of Morocco study also appraised the economic profitability or efficiency of the activities in the sample by calculating their domestic resource cost of foreign exchange (DRC ratio). An activity is economically profitable if its DRC is less than unity. The central conclusion was that there was a close correlation between the DRCs and EPCs. This correlation was particularly close in the case of exports for which bothDRCs and EPCs were less than unity, but somewhat less on the domestic market where DRCs of less than unity were sometimes associated with EPCs that exceeded it significantly.

Table 5.3 Net Effective Protection Coefficients by Market

Sector	---Effective protection--- Domestic market	Export	---Effective subsidy--- Domestic market	Export
Textile & leather	1.05	0.76	1.08	0.79
Mechanical & electrical	1.17	0.75	1.02	0.76
Chemical	1.06	0.46	0.91	0.45
Agro-industry	0.97	0.76	0.95	0.72
Average	*1.08*	*0.72*	*1.03*	*0.66*

Source: Same as Table 5.1.

Consumers and the government, who bear the cost of the high EPCs, were subsidizing economically inefficient activities, some of which were not even financially profitable for their shareholders. Conversely, export industries entail no cost for the Moroccan consumer, make in general a positive budgetary contribution, and tend to be more labor intensive. Despite these desirable attributes, unit value added margins, measured in domestic market prices, were generally lower for exports than for import substitutes—indicating relative anti-export bias. Moreover, absolute anti-export bias existed in that these market price, unit value added margins were often less than in border prices, in sharp contrast to the case for import substitutes. Similar findings resulted from the analysis of incentives of the agricultural sector.[8]

---

8. Estimates of effective protection coefficients for 1984-86 are summarized below:

Type of Agriculture	Import Substitutes	Exports
Rainfed	0.8-1.0	<1.0
Irrigated	1.5	0.8

Note: EPCs varied substantially between markets and areas, irrigated or rainfed. There is price and nonprice anti-export bias. Positive effective protection of import substitution has raised farm incomes in irrigated perimeters. Yet farmers of rainfed areas are comparatively penalized by poorer access to input subsidies and depressed producer prices attributable to the overvalued exchange rate *and* consumer subsidies (flour, edible oils). DRCs for most rainfed crops (cereals, pulses, oilseeds, olives) and export crops (citrus, cotton, potatoes, tomatoes) were less than unity but greater than unity for sugar, which is mainly produced on irrigated land.

NONPRICE INCENTIVES AND DISINCENTIVES TO EXPORTS. The preceding analysis captured neither the inadequacy of some export incentives, nor the nebulous nature of certain other impediments to exports.

First, the profits tax abatement had several deficiencies. Not all direct exports were covered by the decree, which was a source of uncertainty for industrial operators. Indirect exports (that is, manufacturers of goods that are incorporated into exports) were excluded as were export trading houses unless their turnover exceeded 10 million dirham.[9]

Second, interviews with firms had revealed the greater risk of exporting than selling in the domestic market, particularly the risk of nonpayment. To compensate for this, the government had set up an export insurance scheme. It did not function well, however. Nominally, exports were insured for up to 90 percent of their face value, but the effective rate of coverage was often as low as 15 percent of the insured value. Moreover, there were no contingency reserves so that appraisal of risk was very conservative. Many exporters viewed the insurance scheme as a disincentive.

Third, export-related customs regimes (the temporary admission and drawback regimes) were not performing satisfactorily. Regarding temporary admission, not all raw materials were eligible. The administration of the system was inefficient, even arbitrary, especially relating to wastage rates. Customs clearance and transit procedures on the import side, including for export-related shipments, were slow and cumbersome. Firms, especially small ones, were often unable to comply with customs regulations. Their raw materials would then be impounded pending resolution of the dispute, with attendant disruption of production schedules and shipments of finished goods.

Finally, the Internal Revenue Service and the Exchange Control Administration replicated, without exactly duplicating, customs regulations in the wastage rate area. As a result, full compliance with the requirements of all three administrations became difficult if not impossible. This constituted a disincentive to exporters. In addition, the Exchange Control Administration required that export licenses (needed for many products including those manufactured using the temporary admission regime) be stamped at Headquarters in Rabat—more than 700 kilometers from some important production centers.

As regards the drawback regime, the main problems related to delays in payment and revision of the rates. Furthermore, products benefiting from a drawback (for example, metal cans for the sardine industry) could not be imported under the temporary admission regime. This effectively isolated the domestic producer of cans from import competition.

Other impediments to exports resulted from the obligation, in principle, to use national transportation companies for air and sea shipments. Waivers could be obtained, but this wasted time and resources and increased costs. As already indicated, some products required an export license often obtainable only if the declared price exceeded an administratively fixed minimum price. There was a tendency for OCE's export documentation for agro-industrial exports to arrive at destination after the merchandise itself, with the result that foreign importers

---

9. Profits tax abatements increase after-tax profits for profitable firms rather than increase unit value added margins for exported products, regardless of a firm's overall profitability. In fact, such abatements may reduce unit value added margins to the extent that firms are encouraged to reduce their price in foreign exchange in order to increase the share of exports in total turnover.

incurred unnecessary demurrage charges. Consequently, some importers shifted their sources of supply away from Morocco. Problems such as these are difficult to capture in a single measure such as the ESC, yet they constituted an extremely important source of anti-export bias in the Moroccan case.

The ITPA2 program refined this analysis considerably. Whereas ITPA1 had identified customs procedures as a major stumbling block to export development, ITPA2 argued that antiquated international trade procedures, rather than just customs clearance, caused unacceptably long port delays. Completing an export contract entails not only the movement of commodities (the export and its related imports) but also access to, and transmission of, information (related to market development as well as shipment insurance, and payment of raw materials and finished goods). The movement of goods and information are indissociable interdependent processes.

Moreover, transmission of information often occurs on outdated documents interfacing with a mix of equally outdated and modern telephone, telex, or computer systems, which probably do not, or cannot, intercommunicate. Consequently, speeding up the movement of goods through customs may require revisions of procedures elsewhere in the information network (the port, the banks, customs agents, shipping agents, or in the firms themselves). Thus, a global view is indispensable to reform of international trade procedures.

This problem was not unique to Morocco. Following the example of many other countries, Morocco should create a National Commission for the Simplification of International Trade Procedures. Represented on the commission would be all the main participants in the international trade process. Its mandate would be to devise a global and workable strategy for modernizing Morocco's international trade procedures and documents in line with international norms.

The Structure of Taxes and Fiscal Incentives

The Ministry of Commerce and Industry (MCI) undertook a thorough analysis of the structure of taxes for 1982 to complement the findings of the incentives study, which had pointed to the existence of substantial fiscal distortions without being able to document them exhaustively. The fiscal study was considered by MCI as an indispensable component of a global approach to incentives reform (Government of Morocco, 1983b, 1983c). The study focused on import levies, internal indirect taxes including excises and consumption subsidies, as well as taxes on wages and profits. It had a dual focus: first on identifying economic distortions and second on the structure of receipts, especially their distribution between protective import levies and trade neutral (that it, nonprotective), internal indirect taxes levied at equal rates on imports and domestic transactions.

IMPORT LEVIES, INTERNAL INDIRECT TAXES, AND CONSUMPTION SUBSIDIES. The two findings were that significant economic distortions did indeed result from all import levies (SIT, customs duties, stamp duty) and internal indirect taxes (the manufacturer's sales tax or TPS, excise taxes, and consumer subsidies). Second, receipts were excessively dependent on protective import levies. A greater proportion of revenues should be raised through trade neutral, internal indirect taxes. On both grounds, there would be a need to gradually lower import levies and to compensate the revenue losses by raising collections from internal indirect taxes (higher rates and a larger base), as well as lowering consumption subsidies. More specific findings on each tax are given below.

The special import tax was found not to be a uniform tax, despite the claims of many of its protagonists. Its effective average incidence was 12.8 percent as against a legal rate of 15 percent on account of exemptions granted to many categories of products (including cereals, fertilizers, insecticides, and other goods destined to the agricultural sector; some books and news print covered under the Vienna convention; certain capital goods destined to the agricultural sector; and fishing and other boats, airplanes, etc.). These exemptions covered about 15.8 percent of imports.

On the other hand, the mining industrial and tourism sectors paid the SIT on all imports without exception. Particular attention was given to capital goods imported by these sectors since it was argued that the SIT offset the revenue losses caused by exempting capital goods from the customs duty under the investment codes. The study showed, however, that only 20 percent of such imports entered under the investment code and would have paid import duties of 10 or 15 percent. The remaining 80 percent were not exempt from the customs duty. As a result, these incurred total import levies of 35 to 40 percent, net of TPS (the internal turnover tax), or 56 to 63 percent, inclusive of TPS. The latter is the relevant rate when the TPS cannot be deducted from collections on sales or otherwise restituted.

The overall conclusion was that the SIT was itself a distortionary import levy that treated the industrial sector rather less favorably than other sectors, in particular agriculture and fishing. The exemption of the SIT granted to certain goods also discouraged their domestic production because the exemption could not be easily extended to raw materials needed to make them.[10] In addition, as will be pointed out later, the SIT was effectively compensated by offsetting consumption subsidies for those wage goods whose prices were administratively fixed (edible oils and sugar in particular). Finally, the evidence indicated that these distortions had been increasing rather decreasing over time.

The structure of import duties from a resource allocation standpoint had been analyzed in the incentives study. It had shown that duties on raw materials were sometimes greater and sometimes less than on competing imports. The fiscal study highlighted the structure of receipts as shown in Table 5.4.

Exemptions are justified, of course, in some cases (for example, regarding raw materials for exports). Also, some of the goods entering duty free (for example, petroleum products) paid quite heavy internal indirect taxes and/or excises. Others benefited from consumption subsidies. In addition, the global analysis of imports by duty rate indicated substantial intersectoral variation of duties. All in all, this analysis confirmed the findings of the incentives study as to the distorted structure of duties.

There were similar findings of major distortions in the area of indirect taxes. First, rates of TPS on imports and domestic sales of the same commodity were often different, thereby affecting the structure of effective protection. This undesirable outcome may be avoided by equating rates on imports and domestic transactions. Second, the average incidences of the TPS—9 percent for imports and 10.6 percent on domestic transactions—were far below the standard rate of 17 percent, on account of pervasive exemptions and preferential rates. Third, overall effective rates of taxation, including the TPS excise taxes and consumption subsidies, were highly dispersed and varied between negative 35 percent and positive 175 percent.

---

10. The Customs Administration attempted to circumvent this problem by granting the exemption on a case by case basis. This proved quite difficult to administer, as the number of petitions increased. Thus, the SIT was effectively a deterrent to backward integration.

Consumption subsidy outlays amounted to about 50 percent of gross receipts and benefited about 19 percent of taxable transactions.

The study concluded that a global reform of indirect taxes was needed based upon the following principles:

- transformation of the TPS into a VAT with three rates set at the same rate on imports and domestic production;
- elimination of the stamp duty to ensure the neutrality of the TPS/VAT;
- gradual lowering of the SIT and recalibration of the VAT and other internal taxes to recover losses;
- tariff reform aimed at a much lower maximum rate and a considerable reduction of the dispersion of rates in order to reduce resource misallocation effects.

Table 5.4 Structure of Import Duties and Receipts (in percent)

	Import duties	Share of imports	Share of receipts
Range	0	58.1	0
	2.5-7.5	3.6	3
	10-30	35.4	74
	>30	2.9	23
Averages			
Total	7	100.0	100.0
Excluding exemptions	16.6	n.a.	n.a.

Source: Government of Morocco (1983c).

These policies were to be accompanied by a much more limited and transparent policy vis-à-vis consumption subsidies, which also needed to be lowered.

DIRECT TAXES. The primary emphasis was on the fact that the export code granted a ten-year partial or total exemption from the 48 percent profits tax on export-related profits. They were subject only to the PSN equal to 10 percent of the profits tax liability levied on all firms including those exempted from the profit tax. The study pointed out that more taxes could be effectively obtained from the export-related sales, without increasing the profits tax burden on domestic sales, by raising the PSN and lowering the profits tax by an equivalent amount. This would also raise more resources from new firms that benefited from similar exemptions under the investment code. Such a move might be highly desirable in the context of an export-led growth strategy.

More generally, the study pointed out that the average and marginal rates of profits tax were relatively high by international standards. Similarly, taxes on wages and profits were sharply progressive, particularly regarding skilled labor and management. These constituted significant handicaps to competitiveness.

FINANCIAL AND PUBLIC SECTOR ISSUES. In addition to price, production, and fiscal incentives, the ITPA programs examined a variety of related financial and public sector issues. In particular, these concerned the need to

- raise public and private sector savings including those originating from emigrant workers;
- increase financial sector savings from the above sources;
- increase the supply of investment and working capital financing for direct and indirect exporters;
- distribute foreign exchange risk more equitably among the final users of foreign exchange, financial intermediaries, and the Treasury;
- tackle the Treasury and public enterprise arrears problems that were causing major problems for economic management.

Many of the measures in the financial area were quite conventional relating to interest rate policy and structure, as well as to the need to improve the quality of financial intermediation by public and private banks, etc. A particularly important finding was that long-term financial savings might be discouraged on account of bank margins on long-term deposits that were lower than on short-term deposits, and perhaps even were negative. This was because of the placement requirement that compelled banks to invest up to 40 percent of their deposits in Treasury or specialized financial institution obligations bearing very low interest rates (4.25 percent in the case of Treasury bills).

Concerning the public sector, most attention was paid to a preliminary analysis of the sources of inefficiency of public sector enterprises on the one hand and the problems of budgetary management on the other. With regard to enterprises, the starting point of the analysis was the need to reduce the transfers they received from the budget through increased efficiency and better pricing policies. As regards budgetary management, the emphasis was on improving the effectiveness of expenditure controls in operational ministries, which in the past had appeared to cut expenditures by reducing cash outlays while in fact maintaining them through arrears accumulation.

Recommendations

The preceding analysis thus clearly supported the contention that Morocco's external disequilibrium reflected not only excess aggregate demand relative to supply but also a distorted structure of incentives. This discriminated against exports and some categories of production for the domestic market, while simultaneously overprotecting production of other import substitutes. As a result, the economy's capacity to earn foreign exchange efficiently was adversely affected. The resolution of these problems then demanded, in addition to stabilization, an adjustment of the incentive structure that sought to eliminate anti-export bias and excessive protection of the domestic market over a time period long enough to allow firms to adjust. The long-run objective was equality of incentives between import substitutes and exports.

It was argued that such an outward-oriented, supply-side approach could improve the current account deficit while simultaneously increasing output and employment both directly and indirectly through the multiplier effect. On the other hand, a stabilization policy based on domestic deflation would have negative repercussions on output and employment.

In proposing the supply-side approach, the ITPA operations sought to increase the domestic supply of foreign exchange from economically efficient import substitution as well as direct and indirect exports. The emphasis was therefore on reducing anti-export bias, improving the efficiency of resource allocation at the microeconomic level, and maximizing the potential for domestic production of intermediate goods for incorporation into exports. Another important objective was to raise the level of domestic and national savings, especially financial savings originating from emigrant workers. On the macroeconomic front, the need for demand restraint was stressed, particularly regarding the budget, as a sine qua non of generating a sustainable improvement in the external current account. The set of policy actions and other measures for attaining these ends are summarized in Table 5.5.

The centerpiece of ITPA1 was indubitably the reform of trade policy and production incentives for exports and the home market. The top priority was to reduce anti-export bias in the structure of incentives and to increase the domestic supply of foreign exchange from economically efficient exports and import substitutes. The key to this is a compensated devaluation in which the exchange rate is devalued and import levies (that is, the SIT, import duties, and the stamp tax) are reduced in a phased manner. In this way, production incentives may be quickly increased for exports and gradually reduced for the domestic market. In other words, trade policy was to shift from tariff and nontariff protection of the domestic market to exchange rate promotion of production of all tradables, whether for export or home consumption. The short-run priority was thus to encourage exports through the compensated devaluation and other measures. Lowering protection of the domestic market, while essential, would require more time and be politically more difficult. The objective could be more easily attained if better incentives for exports had already increased the potential profitability of exporting compared with the domestic market. Consequently, the tariff and nontariff protection of the domestic market should be phased over a five-year period. This would give hermetically protected firms time to adjust. The key elements would be a phased import liberalization program and a gradual reduction of the maximum rate of import duty.

In turn, the reduction of import levies (that is, the SIT and customs duty) would entail a loss of budgetary revenues and so an increase in the budget deficit. Therefore, other sources of revenues were needed. These were to come from increases in internal indirect taxes (the TPS and excises), from larger budgetary transfers from OCP and other exports where Morocco might have market power, and from the expected growth of exports. The budgetary situation could be further improved by containing unnecessary expenditures. Thus, to be successful, the incentives reform had to be accompanied by complementary measures on the budgetary side.

A final consideration in the design of ITPA1 was to minimize the social cost of adjustment. A particular virtue of compensating the devaluation by a reduction in import levies was that it minimized the impact of the devaluation on consumer prices, and hence on the real wage in domestic prices. At the same time, the devaluation had a full downward impact on the real wage, expressed in foreign exchange, which therefore increased external competitiveness.

This package of recommendations was made only after two alternatives, or some combination thereof, had been rejected. The first was to leave the SIT unchanged and to reduce existing anti-export bias via faster import liberalization and import duty reduction at the existing or devalued exchange rate. This was

Table 5.5 Issues, Objectives, and Corresponding Policy Measures in the ITPA1 Program

Issues and Objectives	Instruments
Remove anti-export bias for direct and indirect exports	Devalue to recapture past appreciation and overvaluation relative to equilibrium exchange rate; manage exchange rate flexibly to sustain competitiveness
	Eliminate export licenses, statistical taxes, and duties on manufactured goods
	Improve customs regimes and international trade procedures
	Improve access to pre- and postshipment working capital finance
Reduce tariff and nontariff protection of domestic market	Liberalize competing imports in a phased manner
	Lower import levies, especially the special import tax, to compensate for the devaluation
	Undertake a multiyear reform of indirect taxes and tariffs hand in hand with the import liberalization program, aiming at a maximum tariff of 25 percent and more uniform effective protection
Minimize adverse impact on budget of reduction of SIT and import duties	Raise transfers from OCP to Treasury in line with devaluation
	Raise internal sales tax, which is not distortionary vis-à-vis international trade
	Reduce government expenditures
Improve domestic regulatory framework to enhance supply response	Remove ex ante price controls, except on subsidized products; remove bias against labor and foreign capital in investment code
	Remove domestic content requirements in certain industries
Encourage mobilization of financial savings especially in the form of time deposits	Ensure that real interest rates are positive
	Remove placement requirements on time deposits

however, to be very difficult to sell politically to domestic manufacturers. The second option was not to devalue, but rather to reduce anti-export bias by the use of export subsidies. Already discussed at length as early as 1980, this option was rejected as being undesirable on economic, budgetary, and fiscal grounds.

The ITPA2 program represented a continuation of ITPA1 in the international trade arena. As noted earlier, it also insisted on the creation of the National Commission for the Simplification of International Trade Procedures. In the financial domain, the ITPA2 program included a component pertaining to public investment and public enterprises. Emphasis was again put on interest rate policy, flexibility, and management, especially reducing public and private borrowing from the banking system and the Central Bank at subsidized interest rates. A mechanism was advocated for reducing the cost to the Treasury of assuming the foreign exchange risk on loans contracted in foreign currency by financial intermediaries for onlending in domestic currency. In the public sector, there were two objectives: to arrest the accumulation of arrears that had resulted from budgetary authorizations exceeding available financial resources and to begin the process of improving the efficiency of public enterprises and reducing their financial dependence on the Treasury. A priority in this connection was to understand how to reduce cross arrears between the Treasury and public enterprises on the one hand and between both of them and the private sector on the other. Finally, regarding macroeconomic management, ITPA2 underlined the need for continued demand restraint in general and in public demand in particular.

Conclusion

It is clear from Table 5.5 and the preceding paragraphs that the ITPA programs, while fundamentally microeconomic in conception, took a very broad view of Morocco's economic problems. The solutions proposed attempted to promote and improve the structure of incentives while minimizing the necessary cost of adjustment on employment and the budget. Thus, a compensated devaluation (that is, one that is offset by a reduction in import levies) has a smaller impact on consumer prices than does an uncompensated devaluation, etc.

In a similar vein, the program advocated recovering part of the revenue shortfalls from lowering import duties via increasing export taxes on phosphates, thereby passing the burden to the rest of the world. Regarding domestic taxation, the programs argued for an improvement in the structure of taxation via a move away from trade taxes as a primary source of revenues and toward nondistortionary internal indirect taxes. On the expenditure side, the programs sought to understand the reasons leading to the accumulation of arrears by the Treasury and public enterprises. And, in the financial sector, the focus was on the key factors that would enhance the mobilization and efficient use of financial savings.

In short, the ITPA programs were the epitome of a focused and detailed supply-side adjustment program with one principal aim, namely, an increase in the domestic supply of foreign exchange from economically efficient exports and import substitution. There were, of course, many secondary objectives, including a conservative fiscal stance without which trade policy reform could not succeed.

Preparation and Negotiation of ITPA

Preparation

Initially, reactions to ITPA1 were mixed. The Ministry of Commerce and Industry (MCI) was a strong supporter, which is hardly surprising given its role in the underlying research and analysis. Its convictions were shared by the Ministry of Economic Affairs, a department of the prime minister's office. In contrast, both the Bank of Morocco and the Ministry of Finance had major reservations about key components. The industrial sector expressed unqualified support only for the simplification of international trade procedures. The International Monetary Fund expressed reservations about the reduction of the SIT and import duties because of the revenue losses. Other bilateral creditors, informally consulted, were broadly supportive. The opinion of foreign commercial banks was not solicited. The initial positions of the main Moroccan participants to the proposed ITPA program will now be considered. Attention will then turn to the ensuing creation of a consensus around a compromise package.

THE MINISTRY OF COMMERCE AND INDUSTRY. Its support emanated from the belief that durable industrial growth required internationally competitive production in terms of price and quality. The economy was to be considered as a whole—not artificially partitioned into import substituting and export-oriented components. Integrated industrial development of capital intermediate and consumer goods could be achieved only by making firms compete internationally at home and abroad. In turn, the rigor of foreign competition would ineluctably create an internationally competitive structure of prices and wages.

In order to compete, Morocco needed to liberalize imports; reduce import duties and their surrogates (that is, the special import tax, the stamp tax, and import duties); pursue a flexible exchange rate policy, devaluing if necessary to make up for past appreciation of the real exchange rate; remove other policy-induced anti-export biases, including administrative impediments to exports; and sharply ameliorate the domestic regulatory framework. In addition, the budget deficit needed to be curtailed, since its existing level could only impede rather than foster economic growth.

The substantial appreciation of the exchange rate between 1973 and 1983 had eroded the international competitiveness of Moroccan goods. From an economic standpoint, foreign exchange could be considered as a basic commodity whose producer price should not be maintained at an unduly low level. An exchange rate adjustment was needed to increase the capacity of the economy to earn foreign exchange, despite the higher local currency cost of servicing existing debt.

However, the ministry argued that the devaluation should be compensated by a reduction in the special import tax and import duties. These complementary measures would attenuate anti-export bias by reducing the difference between unit value added margins in domestic and export markets. Another reason for reducing the SIT was that the Ministry of Agriculture had persuaded the Ministry of Finance to suspend all import levies as well as indirect taxes on many directly and indirectly imported agricultural inputs. The Ministry of Commerce and Industry considered that it would be better to eliminate the SIT rather than suspend it selectively. Forgone revenues could be recovered by an increase in trade-neutral indirect taxes effectively levied on consumption goods only, as suggested by the fiscal analysis undertaken in the preparation of ITPA.

A variety of arguments were advanced in favor of import liberalization. In the first place, the ultimate beneficiary of industrialization should be the consumer rather than firms' stockholders and employees. Industrial sector policy should induce firms to provide goods to national consumers at world prices or lower, after appropriate adjustment for any duly documented excess costs of domestically produced goods and tariffs on imported raw materials. This would automatically provide consumers with a net benefit relative to the alternative of importing. The government should not reserve the domestic market for domestic producers. They, just like exporters, had to win their markets away from competing foreign suppliers with products of the desired quality, competitively priced vis-à-vis imports. Under these conditions, wholesalers could be expected to prefer Moroccan-made goods.

Second, recent experience, especially since March 1983, had demonstrated the difficulties and costs for the administration and firms alike of managing universal import restrictions equitably. These problems were compounded by the need to centralize the process in Rabat. Furthermore, the creation of rents for import license holders was undesirable in itself and was also one of the factors generating demands for rents elsewhere in the economy.

The MCI's third argument for minimizing import restrictions related to investment capacity utilization and antitrust policy. Free trade (that is, the absence of nontariff barriers and prohibitive duties) provided a natural disincentive to wasteful investment. In contrast, hermetic protection could lead to domestic monopolies or oligopolies. Initially high profits, resulting from collusive behavior, could entail a buildup of excessive capacity compared with demand. In turn, chronic excess capacity could lead to predatory pricing and financial losses, and the associated financial and social difficulties of managing enterprise restructuring and closedowns. These were often resisted not only by labor but also by the banks that had financed the original investments.

Moreover, the creation of excess capacity could even result in negative import substitution (that is, the foreign exchange cost of investment goods exceeds the foreign exchange savings generated by production over the project's life). This undesirable outcome could be avoided only by administrative authorization of all domestic investment. But this would be complicated, undesirable, and politically unacceptable. Hence, it was better to avoid the problem altogether by eschewing restrictions on competing imports.

At the same time, the Ministry of Commerce and Industry announced that it would always be ready to intervene to "protect" firms against unfair foreign competition, especially dumping, in the context of a public adjucatory process. Categorical affirmation of unfair competition, without proof or reasoned argument, would not be acceptable. Consequently, the MCI actively supported the creation of an interministerial consultative committee on protection that would be empowered to make rulings in this area. The committee's terms of reference would call for it to intervene in order to safeguard existing industries against unfair foreign competition rather than to protect domestic industries against all imports. The preferred means of intervention should take the form of antidumping tariffs. They would be levied to the extent that it could be shown domestic industries had suffered "material injury" along the lines outlined by GATT agreements.[11]

---

11. The incentive research team had spent considerable time preparing written guidelines for the evaluation of petitions for protection. Updated during the preparation of ITPA1, these guidelines were to be one of the instruments used by the interministerial committee. This kind

THE MINISTRY OF FINANCE. Initially, the only ITPA1 recommendations supported by the Ministry of Finance concerned the need to simplify international trade procedures as well as the structure of import duties, which were becoming very difficult to manage. The Ministry of Finance believed that all other ITPA1 measures would aggravate rather than improve the current account and budget deficits. Import liberalization would certainly cause competing and noncompeting imports to rise. Reduced protection of the domestic market would jeopardize the survival of Morocco's inexperienced industrialists. At best, a devaluation would have a marginal impact on industrial exports because of the low elasticity of supply in Morocco and import restrictions applied by Morocco's trading partners, particularly in the European Community. The overall result would then be a deterioration rather than an improvement in the current account deficit.

The budget deficit would be aggravated because of the revenue loss from the reduction of import duties. This could not be compensated by an immediate increase in the rates of the TPS, which was shortly to be transformed into the value added tax. In any event, raising more revenues through raising rates of either tax would be very difficult given the recent increases in the standard rate of the TPS.[12] Moreover, increasing preferential rates of TPS applied to basic consumer essentials was ruled out on sociopolitical grounds. In addition, it was felt that the devaluation would not enable OCP to enlarge its transfers to the budget, as the Ministry of Commerce and Industry had implied, because of the cash requirements of its impending expansion.

The Ministry of Finance also objected to the proposed program on numerous other grounds. First, a devaluation would entail a rise in consumption subsidies since the prices of flour, sugar, edible oils, and butter could not be further raised for political reasons. Second, the cost of servicing the external public debt would also be augmented. The program would not increase government revenues because export-related revenues were, to all intents and purposes, exempted from income taxation under the export code. Only marginal increments in tax revenue would be generated from additional employment that was concentrated in low-wage employment.

Raising interest rates would lower investment and so should be avoided. It was not necessary to increase the already sufficient supply of subsidized Central Bank prefinancing to exporters. Indeed, in some cases this had been fraudulently diverted to finance production destined to the domestic market where working capital loans paid higher unsubsidized rates of interest. Extending preferential rates to indirect exporters was not appropriate and would be administratively very complicated.

Finally, Morocco was in the midst of negotiating a new IMF standby that incorporated as a performance criterion an important reduction of the Treasury deficit. The standby was in direct conflict with the ITPA program unless the International Monetary Fund would accept a larger Treasury deficit to accommodate the World Bank. Also, past increases in import levies, particularly in the SIT, had met with IMF approval, so why reduce it now?

---

of institution building was perceived to be a particularly valuable external effect of the incentives research project.

12. The standard rate had been increased from 15 to 17 percent in the 1982 budget and to 19 percent in July 1983.

THE CENTRAL BANK. The Bank of Morocco shared the Ministry of Finance's reservations about the proposed ITPA1 program, particularly regarding the current account and budget deficits.

THE MINISTRY OF ECONOMIC AFFAIRS. This ministry was a strong advocate of ITPA1, especially its orientation toward growth rather than stabilization. Three aspects of the proposed program were particularly desirable:

- the attempt, through exchange rate policy, to increase earnings not only from exports but also from tourism and worker remittances;
- the liberalization of imports to compel domestic firms to become more efficient;
- the use of a compensated devaluation, which minimized price increases on the domestic market;
- the expansion of labor-intensive exports, which would increase employment.

The ministry attached particular importance to the economic and political desirability of employment generation, which could counterbalance the negative effects of higher prices that might arise from the program.

THE INDUSTRIAL SECTOR. Import substituting firms that produced solely for the domestic market were alarmed at the probable consequences of import liberalization. Hermetic tariff and nontariff protection to foster industrialization had been the government's policy since 1973 and to change it in midstream could spell ruin for many firms. Moreover, new investment would be discouraged by a free trade policy.

REACTION TO THE INDUSTRIAL SECTOR ARGUMENTS. Industrialists in Morocco, particularly those producing for the domestic market, are a powerful lobby. Their reaction to the ITPA1 program provoked considerable debate that in turn prompted further arguments for and against it, as described below.

Both the MCI and the Customs Administration pointed to the practical difficulties of administering tailor-made protection for all goods and all firms. There were always deserving exceptions. One exception of particular importance concerned raw materials used by exporters. In the past, the right to import duty free had been subject to prior authorization by the Customs Administration and the MCI in order to compel exporters to maximize backward linkages. This policy, which was really equivalent to an import license, had penalized exports. Accordingly, a new policy was needed that would grant direct exporters, as well as their domestic suppliers, the right to import duty free without prior authorization. In this manner, exporters could be encouraged to maximize backward linkage through the use of domestically produced raw materials that would be internationally competitive in terms of price and quality.

In a similar vein, prohibitive tariffs were inequitable because it was impossible to define tariff categories with sufficient detail to protect the production of one particular firm. As a result, other "nontargeted" imports would have to pay the tariff. Unavoidably, their user's costs and competitiveness would be affected adversely. In addition, some prohibitive tariffs were unnecessarily high and greater than the difference between the domestic ex factory and international prices. Under these conditions, the protective tariff could be somewhat reduced without harming domestic producers.

Moreover, the MCI's intent was not to remove import licenses and reduce tariffs in a brutal manner. On the contrary, it proposed to accompany the liberalization of imports with an increase or a reduction in the tariff to a level that would initially make firms competitive vis-à-vis imports at existing prices. Under these conditions, the evolution of domestic sales would depend on the ability of domestic firms to compete in terms of quality. Efficiency gains would subsequently be spurred by phased reductions in tariffs.

The MCI emphasized that the tariff would be determined with industrialists in a collaborative yet objective manner. The ministry—after a study on the basis of firms' production cost data—would determine a tariff based on the assumption that capacity would be fully utilized. Firms would have the right to comment on the study and to make counterproposals. Adequate antidumping safeguards would be proposed if there was a high risk of international predatory pricing. Finally, the MCI assured firms of its willingness, in the event of unexpectedly catastrophic results, to reintroduce import restrictions on an emergency basis since its role was to support and promote economically efficient domestic industry.

The MCI also advanced other arguments. One could not expect a sector to accept import liberalization easily if its own suppliers remained protected. Hence, the approach would be to liberalize targeted industries and their suppliers simultaneously, yet in an orderly manner over several years. Second, deprotected firms would profit from the devaluation and reform of export incentives to sell more overseas. Finally, the ministry felt strongly that hermetic protection had outlived its usefulness. Its maintenance was perennially masking other serious handicaps to industrialization—for example, price control at the producer and consumer level, internal and external transport, access to credit by small-scale enterprises, and legislation pertaining to employment, bankruptcy, quality control, and manpower training.

The Negotiated Program

The ITPA program was first presented to the Government of Morocco just after the emergence of the March 1983 foreign exchange crisis. At this juncture, discussions were already under way concerning a new IMF standby that aimed at reducing the current account and budget deficits. Moreover, foreign commercial banks were by now unwilling to increase their exposure to Morocco. Bilateral assistance was also less forthcoming. The result was a large shortfall of external resources relative to needs. This combination of circumstances led Morocco to attempt a stabilization cum adjustment program, the former being financed by the IMF and the latter by the World Bank. The realization of the program thus required a consensus between the World Bank and the IMF on the one hand and among the different ministries on the other.

The ITPA loan (US$150 million) was finally negotiated in November 1983. It concentrated on the following themes (enumerated in Table 5.5):

- removal of anti-export bias for direct and indirect exports through a compensated devaluation and improved export incentives;
- enhanced competitiveness of import substituting firms, vis-à-vis imports, in terms of price and quality via a phased import liberalization and tariff reduction;
- compensation of revenue loss for the Treasury by raising internal indirect taxes and transfers from OCP;

- amelioration of the regulatory framework by liberalization of price control, domestic content requirements, reduction of restrictions on foreign investment, etc.;
- enhancement of savings and their financial mobilization through a more appropriate interest rate policy.

Specific additional points with regard to each theme are noted below.

In the first place, it was agreed that the SIT would be reduced in three equal, annual installments beginning on January 1, 1984—whereas the original proposition had been to eliminate it in one go. As a result, the devaluation of 17 percent was initially only partially compensated. Second, in light of the forthcoming tax reform, no immediate increases were planned in internal indirect taxes over and above the two-point upward adjustment of the standard rate that had occurred in July 1983 as part of the IMF program. Instead, the IMF agreed that the budget deficit would be revised upward by the amount of the SIT revenue shortfall, with ITPA's counterpart funds being used to provide the corresponding financing. It was expected that the additional receipts would be procured through introduction of the VAT in the tax reform process. Third, no measures were taken to enhance the mobilization of financial savings, in particular by reducing the share of time deposits (30 percent) that had to be invested in Treasury bills.

As regards import liberalization, the MCI insisted on announcing each phase of the program well ahead of time so that firms would have time to prepare for the adjustment. However, the ministry rejected the notion of coupling import liberalization with firm-level restructuring programs. It feared that import liberalization might become hostage to the adoption of such programs and so preferred to expose firms to foreign competition behind an appropriately determined tariff rather than await the design and implementation of a restructuring program.[13]

Sometimes import liberalization was accompanied by the use of reference prices to provide additional protection. This was achieved by calculating import duties and taxes on the basis of the reference price rather the declared C&F value. The MCI did not favor the use of reference prices, but it had to concede to politically powerful pressure groups in the industrial sector. Nevertheless, these reference prices were fixed in domestic currency so that their foreign exchange value, and hence their protective effect, would decline over time with the expected depreciation of the dirham.[14]

Finally, it had to be determined how ITPA's foreign exchange resources might be used. Some considered that all imports except luxury goods could be financed, whereas others favored the creation of an export development fund that would finance only raw material imports required by exporters. This second proposition was strongly resisted by the Ministry of Finance, which contended that exporters received, in a timely manner, all the foreign exchange they needed. Accordingly, the first option of general import financing was retained.

---

13. Thus, in the ministry's view, restructuring problems should be dealt with as the need arose after liberalization.
14. The Office des Changes also opposed reference prices as a potential source of overinvoicing.

Reaction to ITPA by the Bank's Board

In general, the ITPA approach and operation was well received and viewed as an example that should be followed in other countries. First, the ITPA approach attempted, via the compensated devaluation and other microeconomic measures, to lay the foundation for economic growth while minimizing the social cost of adjustment, particularly as regards prices. The second desirable characteristic was to have worked with the IMF so as to create a reform package that combined growth-oriented microeconomic adjustment with macroeconomic, especially budgetary, stabilization. Thus, an indispensable macroeconomic dimension had been introduced to the microeconomics of adjustment. Morocco's problems could be resolved only by reducing the budget deficit and by improving the efficiency of resource allocation at the microeconomic level. Neither action alone was sufficient. Third, ITPA represented a focused approach based on joint analysis undertaken in Morocco by the World Bank and the Moroccan government. It had successfully avoided the diffusion and overambitious comprehensiveness of past structural adjustment programs.

There were, of course, arguments against the project. The most important reservation was that the Bank was counseling Morocco to liberalize imports and reduce tariffs when the country could least afford it, given the magnitude of the external current account and Treasury deficits. In this view, more effort should have been directed toward raising savings through expenditure reduction, particularly by the government.

This argument was countered by pointing out that savings could be raised by cutting expenditures or raising incomes, or both, as was needed in Morocco. Short-run public expenditure reduction was considered one of the main objectives of the IMF stabilization program. ITPA's objective was to promote larger private savings from GDP growth of economically efficient exports and import substitution. Import liberalization had to be seen in this context.

Specifically, higher growth would require higher rather than lower import volumes as well as more efficient use thereof in economic terms. In this regard, import liberalization was to be seen as a means of compelling import substituting firms to become more competitive vis-à-vis foreign products in terms of price and quality.[15] However, the import liberalization program was pluri-annual and would be phased no faster than the rate of growth of exports and available foreign exchange financing. Also, the rate of import liberalization would be conditioned, in part, by the evolution of public and private expenditure.

In response to the criticism that the reduction of the SIT would increase the budget deficit, it was pointed out that the original proposal had been to recover the revenue loss by increasing the TPS as well as OCP transfers to the Treasury, and by cutting expenditures. The choice of specific measures had been left to the government and could be phased in, with tax reform being undertaken in the context of the IMF standby program. The only loan requirement was that the SIT should not be compensated by an increase in customs duties or other nonneutral trade taxes. In this regard, the government had decided, with IMF approval, not to increase the TPS over and above the two-point increase in the standard rate that had occurred in July 1983 as part of the IMF standby. Instead, the 1984 budget deficit

---

15. Import liberalization also could be a means of preventing wasteful investments in import substitution where some industries had entailed negative net foreign exchange earnings for Morocco.

would be allowed to increase by the amount of the SIT reduction. In turn, it had been agreed that ITPA's counterpart funds would be used to make up for the revenue shortfall. However, this was only a stopgap measure. In the longer run, the ITPA program could be successful only if accompanied by a successful reform of indirect taxes, which was the IMF's responsibility.

Another argument against ITPA was that it moved the Bank away from its traditional activity of investment lending. In response to this concern, it was pointed out that the correction of the policy framework and incentive structure was a sine qua non of a higher volume of economically sound investment lending.

Discussion and Preparation of ITPA2

As indicated earlier, ITPA2 represented a continuation of ITPA1. It was prepared between late 1984 and mid-1985. Discussions of ITPA2 within Morocco and the World Bank were broadly similar to those concerning ITPA1, but they took place against a different background: the picture regarding the budget deficit and public sector arrears was much bleaker. As a result, there was a much sharper focus on how to recover the revenue shortfall resulting from the reduction of the SIT. In broad terms, the MCI and the World Bank pressed for a compensating increase in the vector of TPS rates. The Ministry of Finance argued, as it had under ITPA1, that this was politically impossible. In the end, their view prevailed. So, in January 1985, the SIT was reduced from 10 to 7.5 percent rather than to 5 percent as initially agreed under ITPA1. ITPA2 provided that this level would be attained in January 1986. Elimination was forecast for January 1987, one year later than originally scheduled.

In all other respects, the ITPA2 program in the area of international trade was in line with the ITPA1 approach: continued exchange rate flexibility, further import liberalization, and reduction in the maximum rate of customs duty to 45 percent. A provisional tariff was also to be prepared that would provide guidance for future stages of the tariff reforms, notably the attainment of a maximum rate of protection of 25 percent by the end of 1988. On the export side, specific agreement was reached that the statistical export tax (0.5 percent) and more export licenses would be eliminated. Further improvements in customs regimes were agreed, particularly concerning their use by indirect exporters. Customs disputes were to be settled in an a posteriori basis rather than the a priori basis used until then. Customs clearance delays were to be reduced by 50 percent.

The National Commission for the Simplification of International Trade Procedures was created. Its mandate was to simplify and modernize all international trade procedures and documents and the treatment of information relating to trade. A new export code was planned that would also extend profit tax incentives to indirect exporters.[16]

In the financial area, agreement was reached on elimination of placement and reserve requirements for time deposits accepted by deposit banks. Success in mobilization of financial savings by commercial banks was to be rewarded by the Central Bank in the form of higher credit ceilings. A foreign exchange risk fund was created in the Treasury with participation of the specialized financial institutions so as to reduce the cost to the Treasury of assuming the foreign exchange risk on the loans they contracted abroad. The Treasury agreed to seek

---

16. Proposals in this area included reducing fiscal incentives to direct exporters pari passu with the depreciation of the exchange rate.

more short-term financing from the money market and to float medium- to long-term bonds. Regarding public investment and public enterprises, new budgetary techniques were introduced to keep expenditure commitments in line with available resources so that arrears accumulation would cease. Finally, a start was made on increasing interbank competition.

Implementation and Evaluation

This section first assesses to what extent the ITPA programs were implemented as agreed. Then it evaluates the programs, particularly regarding

- increasing the supply of foreign exchange from economically efficient exports and import substitution;
- raising the level of domestic and national financial savings;
- the evolution of the current deficit and the budget deficit;
- microeconomic developments.

The discussion will draw primarily on the methodology developed in Chapter 3 as well as on a recently completed study of the impact of the program on trade and industrial adjustment (World Bank, 1988a).

With one significant and several minor exceptions, the ITPA programs have been well implemented, though not always without difficulty and delay. Areas of less than successful implementation concern the reduction of the SIT; recovery of forgone revenue through raising the TPS/VAT; higher revenues from OCP; the extension of fiscal and financial incentives to indirect exporters; and the implementation of a global approach to simplification and modernization of international trade procedures.

Areas of Successful Implementation

EXCHANGE RATE POLICY. A flexible exchange policy was effectively followed beginning in mid-1983, with the trade-weighted real exchange rate depreciating about 25 percent. In fact, most of the adjustments had occurred between September 1983 and late 1985. From 1986 to 1987, the exchange rate was effectively stable, and in 1988 it slightly appreciated. During 1984 and 1985, the Central Bank continued to object that exchange rate depreciation led to unnecessary increases in the cost of debt service and consumption subsidies without producing commensurate benefits in the form of higher exports and lower imports. Indeed, in 1984 and 1985, exports had risen only marginally in the aggregate, whereas imports had shown much sharper growth. As will be discussed later, these trends were reversed in the next two years.

IMPORT AND EXPORT LICENSING. Import licenses were effectively removed, sometimes for more products than originally negotiated, and generally in a timely manner. By the end of 1986, prohibitions had been abolished, and import licensing applied to only 13.7 percent of import value compared with 61.3 percent in 1983.

The key to this success has been the tariff policy that accompanied the removal of the import licensing. As far as possible, the government kept to its word: it informed the industrial sector of which products would be liberalized in any given year and allowed it to petition for adjustments in tariff rates within the ceiling rate. For a limited number of products, additional protection was given through reference prices. These prices were not used, however, as a nontariff barrier to replace import licenses, but rather as the customs valuation basis for imports whose values seemed abnormally low. Used in this manner, the reference prices correspond to a specific import levy (equal to the ad valorem rate applied to the reference price). Despite considerable pressure, reference prices have been used on only a few products. In general, implementation delays occurred only for products whose licenses were controlled by ministries, other than the Ministry of Commerce and Industry, that had not been a party to the original program. In most cases, the difficulties were resolved.

On the export side, licenses were removed on schedule although sometimes with great difficulty (hides and skins, mining products). Similarly, OCE regulation of exports of agro-industrial exports and processed fish (canned sardines) was removed with the abolition of licensing and minimum prices. Subsequently, there were attempts at retrenchment, some of which have succeeded on account of noncompetitive collusion among producers rather than because of government intervention. For example, the fish canning industry was initially in favor of full liberalization but subsequently attempted to reintroduce industry-controlled minimum prices. This partly reflected the failure of an old, excessively fragmented industry to restructure rapidly enough in the wake of the liberalization. At the same time, the reversal raises the interesting and important question of whether complete price liberalization was correct, given that some canned fish enter duty free, under a quota, into EC markets. Perhaps a more appropriate policy would have been to advocate the use of an export tax or an auction of the quotas for EC markets.

Similarly, there has been strong pressure to reintroduce export licenses on hides and skins in order to prevent these exports from depriving leather manufacturers of badly needed raw materials. To date, these arguments have been countered by pointing to the fact that imports are unrestricted and that it is not desirable on economic or equity grounds to deprive hides and skins exporters of their preferred markets. Similar arguments have been advanced against recent proposals to introduce export licenses for other tradable agricultural and marine raw materials that have been requested, in order to guarantee an adequate supply of raw materials to domestic processors who wish to increase their exports. These processors argued that their petitions should be granted simply because of the additional value added created at the processing stage, without consideration of the reduction in value added that would result for the raw material exporters. There is an important lesson here—namely, that the sustainability of external deregulation necessitates sustained effort on the part of dedicated civil servants and politicians who are prepared to maintain their positions in the face of relentless bureaucratic and political pressures.

IMPORT AND EXPORT DUTIES. In this area, ITPA's contractual obligations were met. The maximum rate of duty was reduced from 400 percent to 45 percent on schedule. Many exemptions were eliminated by the introduction of a minimum import duty of 2.5 percent in January 1987. Average tariffs had also been considerably reduced from 36.1 to 23.4 percent. Dispersion, as measured by standard deviations, has decreased from 112 percent to 65.8 percent of the mean.

Yet, in certain respects, it could be argued that the measures actually implemented fell far short of what had been initially planned, even though these plans considerably exceeded what was legally required to comply with ITPA conditionalities. Specifically, it was originally intended that protection would not exceed 25 percent by the end of 1988. This target, whether interpreted in nominal or effective terms, was not met. The tariff remains exceedingly complicated, with over 8,000 positions, and is in urgent need of simplification. This has become all the more clear since the preparation of the Harmonized Tariff.

Nevertheless, in retrospect, it may be argued that the approach to tariff policy has been relatively sound, in particular the emphasis on removing as many quantitative restrictions as possible before reducing the maximum tariff below 45 percent. In this way, it has been possible to avoid a situation in which import liberalization and tariff reduction become contradictory policies. A tariff cap of 45 percent, however, remains high and should be reduced as fast as possible. On the export side, the statistical export tax (0.5 percent) was removed on schedule.

SPECIAL CUSTOMS REGIMES. The temporary admission regime and drawback regimes were substantially improved for both direct and indirect exports. The Customs Administration was particularly innovative in allowing exporters to declare their own wastage rates and in accepting the principle of ex post verification (by the administration within six months of exporting). The Customs Administration also introduced a mutual guarantee system for goods imported duty free under bond. Under this system an exporter could replace the traditional bank guarantee, which usually cost 2 percent per year, with a guarantee from another firm. As a result, firms were able to obtain more working capital credits from commercial banks. These measures were especially well received by exporters.

Nevertheless, substantial problems remain regarding the temporary admission system. In particular, there was no improvement in the accounting procedures used by the Customs Administration to reconcile reexports of goods imported under the temporary admission regime. This issue requires collaboration by several different parties—the Customs Administration, the organization providing the guarantee, and the exporter. At the time of the ITPA operation, the importance of this collaborative approach was not yet fully appreciated. The issue was subsequently taken up in 1987, but even then a satisfactory approach was not found. Today the issue remains as important as ever. More will be said on this in the discussion of international trade procedures.

INFORMATION CAMPAIGN. This crucial aspect of the program was implemented as agreed. But more important was the "open" nature of the reform program in which the government made it clear to the industrial sector that the intent was to undertake sustainable reforms. Open exchanges of view were regularly sought and encouraged. The timing of the import liberalization and the determination of import duties were discussed with firms, within the context of the government's contractual obligations to the World Bank.

PRICE LIBERALIZATION. Here, again, there were few if any implementation difficulties. Price controls were retained only for subsidized basic products, public utilities, and commodities for which there was only limited competition. For all other products, the principle of self-revision was adopted: firms were able to freely revise their own prices. But the price control authorities retained the right to mandate a reduction, if necessary. This right has never been exercised. Ongoing evaluation of price deregulation by the Ministry of Economic Affairs suggests that it has not had significant adverse effects.

FINANCIAL SECTOR. The main objective of raising financial savings was effectively pursued through the elimination of obligatory and reserve requirements on savings and term deposits. Similarly, an incentive to generate deposits was provided to the banks by determining their credit ceilings in proportion to the amount of deposits effectively mobilized. Interest rates were raised to significantly positive real levels. The budget reduced its reliance on financing at subsidized rates through the obligatory placement requirements and increased recourse to borrowing at market rates from the public and the commercial banks. However, this process was constrained by the size of the Treasury deficit, which did not permit the government to forgo all of the resources from its subsidized borrowing. As agreed, a foreign exchange risk coverage fund was created in order to reduce the cost to the Treasury of servicing foreign debts contracted by financial intermediaries.

PUBLIC ENTERPRISE AND PUBLIC INVESTMENT. The most important measure in this regard was the introduction of revised budgetary procedures to ensure the authorization of expenditures only for which resources were effectively available. A start was made on public enterprise reform and the crucial problems of their arrears. Much more detailed work was substantially undertaken in the context of a public enterprise restructuring loan (PERL) that was negotiated in 1987.

Areas of Less Than Successful Implementation

EXTENSION OF FISCAL AND FINANCIAL INCENTIVES TO INDIRECT EXPORTERS. One of ITPA's objectives had been to maximize the domestic value added content of exports through an extension of customs regimes, as well as fiscal and financial incentives, to indirect exporters. Although this worked well for customs regimes, real progress, as distinct from formal compliance, was virtually nonexistent for fiscal and financial incentives. On both fronts, there was substantial opposition from the Ministry of Finance and Bank Al Maghrib. Both sets of measures, they argued, would be administratively difficult and unnecessarily costly for the budget. These efforts were pursued in the subsequent Industrial Finance Operation, again without success.

Regarding fiscal incentives, the initial objective had been to revise the export code so as to extend to indirect exporters the corporation tax exemption granted direct exporters on their export-related income. Specifically, they would have been granted a reduction in their income tax liability prorated to their sales delivered under the auspices of the temporary admission regime. Simultaneously, the rate of the corporation tax would have been lowered and the National Solidarity Levy (PSN) would have been raised so that the total tax direct burden on profits of firms producing only for the domestic market would have stayed unchanged. But, for direct exports, the effective rate of taxation would have risen pari passu with the increase of the PSN. And for indirect exports, it would have fallen to the rate implied by the PSN. At the end of the tax holiday, the rates would have risen to the normal rate of corporate tax plus the PSN. In one version of this proposition, the tax holiday would have been permanent for exports, but the PSN would have been raised to the long-run desired rate of the corporation tax.[17]

---

17. Thus, exporters would have become effectively subject to the long-run desired rate of the corporation tax, though under the guise of the PSN.

No consensus, however, could be reached on these propositions. A reform of the export code was adopted in which a higher rate of PSN was used during the tax holiday, the length of which was cut from ten to five years. But the extension to indirect exporters did not occur, as the MCI's economic and fiscal arguments were unable to prevail over those advanced by the Ministry of Finance.

Regarding indirect taxes, the idea was to grant indirect exporters the same right as direct exporters to affect their domestic purchases, net of TPS, up to the amount of sales, also delivered net of TPS, to exporters in the preceding year. This change never materialized, with the result that certain indirect exporters, especially manufacturers of packaging who deliver to exports outside the TPS/VAT system, find themselves systematically in a tax credit position vis-à-vis the Tax Administration and thus are obliged to petition and wait for restitution. This has an undesirable impact on the cost of their goods and hence the external competitiveness of Moroccan products. In attempting to solve this problem, which remains significant, another old problem again came to light—namely, that the Internal Tax and Customs administrations have substantially different legislation and procedures regarding wastage rates and reconciliation of imports and reexports of raw materials entering under special customs regimes. As a result, there is a risk that exporters, direct as well as indirect, may find themselves in infraction relative to one or the other. This also is an issue where much work remains to be done to minimize such problems for exporters. The present state of affairs is likely to persist so long as administration of the TPS/VAT remains divided between the two administrations.

Regarding financial incentives, there was considerable discussion during ITPA preparation as to how to extend subsidized Central Bank prefinancing to indirect exporters. As indicated, this was strongly resisted by Bank Al Maghrib and the Ministry of Finance on the grounds of budgetary cost, administrative complexity, and diversion of funds to the domestic market at a time when credit expansion had to be severely limited. Nothing further was done in the ITPA programs, but in subsequent operations this topic received considerable attention. However, the discussion focused less on the provision of subsidized credit and more on the question of whether exporters, direct and indirect, were getting enough prefinancing working capital from the banking system as a whole and whether there was scope in Morocco for the use, as in Korea, of a domestic letter of credit system. Here again discussions have proved inconclusive.

GLOBAL APPROACH TO TRADE SIMPLIFICATION. Progress was made under ITPA1 and 2 regarding customs clearance times, which have been reduced from ten to five days on the import side. But progress has fallen short of what might have been expected because getting the National Commission for Simplification of Trade Procedures off the ground proved to be difficult. Initially, it will be remembered, the incentives research project and ITPA had sought to remove administrative delays occurring in Customs and the Exchange Control Department. However, during ITPA2, it became clear that the earlier analysis was partial at best since inefficiencies existed not only in Customs and the Exchange Control Department, but also in the port, the banks, shipping companies, freight forwarders, and the exporters themselves. A coalition for reform therefore had to be built among all of these parties around a common theme.

These ideas did not really mature until after the reduction of the second tranche of ITPA2 in late 1986 and early 1987, and until after the World Bank had completed analysis of port management problems. This analysis indicated that a series of factors contributed to major port clearance delays that could attain twenty days on the import side. These transit times could be reduced only by a global approach to the simplification of international trade procedures. The starting point for such an approach is that an international trade transaction consists of the processing and transport not only of goods but also of information. The ITPA program had concerned itself only with the cost of manufacturing goods—completely ignoring problems relating to (a) their transport and (b) the processing and transport of trade-related information. Once this point was understood, it became clear that delays to moving goods was the result of inefficiencies in the processing and transport of all trade-related information. The Trade Simplification Commission then began to work—albeit very slowly—around three themes: (1) reduction of port, and not just customs, clearance delays of 50 percent in two years, (2) adoption of import and entry forms aligned with international norms (UN Layout key), and (3) the computerization of the ports and customs to permit computer-to-computer transmission of trade data, both nationally and internationally.

THE REDUCTION OF THE SPECIAL IMPORT TAX. This is the one area where there was formal noncompliance with the ITPA program. As may be seen from Table 5.6, the pace of reduction of the SIT was slowed down beginning in 1985. In 1987, the reduction of the SIT from 7.5 to 5 percent was offset by an equivalent increase in all import duties up to and including 42.5 percent. The minimum import duty rose from zero to 2.5 percent, but was not applied to goods exempted under the investment code or under the auspices of other special legislation (for example, regarding agricultural inputs, boats and airplanes, etc.). In January 1988, the SIT was formally abolished, together with the stamp duty. They were replaced by a fiscal import levy—prélèvement fiscal à l'importation (PFI)—of 12.5 percent. All exemptions from the SIT have been carried over to the PFI.

Taking account of the 2.5 percent minimum import duty, any nonpreferential import now pays customs levies, including the PFI, of between 15 percent and 57.5 percent excluding VAT—compared with a range of zero to 45 percent if the original ITPA program had been adhered to.[18] Import levies, net of internal indirect taxes, are thus significantly higher than originally intended. And, unfortunately, there has been an increase in tariff protection granted to the domestic market and a corresponding increase in anti-export bias.

This sequence of events resulted from a clash between the exigencies of adjustment and those of budgetary stabilization. On the adjustment side, the ITPA program has been carefully built up on the basis of experience in 1982 and 1983, when there had been intense debate as to whether revenues should be raised by further increases in the SIT or through adjustments of the TPS. The former had been advocated by the Ministry of Commerce and Industry and the latter by the Ministry of Finance. On both occasions, however, the Ministry of Finance had accepted the position of MCI that raising the TPS was preferable (because it had a larger base that included not only imports but also domestic production and because it reduced anti-export bias). Both increases had also been submitted to and approved by Parliament. On these grounds, it was felt that a reduction in the SIT could be envisaged, if accompanied by the complementary adjustment of the TPS rates. But these should concern the vector of rates rather than just the standard rate as had

---

18. This last calculation supposes that the stamp duty would have been abolished. Had it been maintained at 10 percent, total import levies, net of internal indirect taxes, would have been between zero and 52.25 percent.

occurred in the past. These adjustments could occur, or not, in the context of the proposed introduction of the VAT, which had been the focus of discussion with the IMF for the past several years. At the same time, the ITPA program was predicated on the realization of increased transfers from OCP to the Treasury in the wake of the devaluation.

Table 5.6 Proposed and Implemented Rates of the Special Import Tax and Stamp Duty, 1983-88

	Special import tax (SIT)		*Stamp duty (SD)*		*Total (SIT or PFI, SD, and minimum import duty)*	
	Proposed	*Actual*	*Proposed*	*Actual*	*Proposed*	*Actual*
1983	15	15	10	10	16.5	16.5
1984	10	10	10	10	11.0	11.0
1985	5	7.5	10	10	5.5	8.25
1986	0	7.5	10	10	0	8.25
1987	0	5.0	10	10	0	8.25
1988	0	0	10	0	0	15.0

There was disappointment on both fronts. Implementation of the VAT was substantially delayed, and the Ministry of Finance was opposed to increasing TPS rates before the reform. Moreover, the reform led to much lower revenues than expected. A similar situation prevailed with respect to transfers from OCP to the Treasury.

From the standpoint of budgetary stabilization, the reduction of the SIT had always been problematic. Opposition to it increased in 1985 and 1986 as the extreme gravity of the public finance situation emerged, particularly regarding arrears and expenditures in consumption subsidies. At the same time, the introduction of the VAT resulted in more rather than fewer exemptions. Net collections showed, at best, marginal growth because rates were not recalibrated to compensate for the generalization of deductions. Extension to wholesalers proved extremely difficult since there was widespread resistance to paying the VAT on existing stocks without the right to deduct the TPS paid at the time of acquisition.

Consequently, the needs of budgetary stabilization prevailed over those of adjustment. The problem is that the current policy is strikingly similar to that of the mid-1970s when the authorities tried to solve the budget deficit by raising the SIT and so increasing protection to the domestic market. Indeed, the Parliamentary debate of late 1987 revealed considerable latent demand for even more protectionist policies through setting the PFI equal to 17.5 percent rather than 12.5 percent. The negative consequences for the dual objectives of reducing anti-export bias and maximizing the domestic content of exported manufactures do not seem to have been appreciated.

Evaluation Strategy

Adjustment programs of the ITPA variety should be evaluated from both macro- and microeconomic standpoints. Especially in the early stages of implementation, it is important to dispose of analytical tools that enable an appreciation of whether things are on the right track. In the Moroccan case, this clearly meant bringing the external current account deficit into line with the reduced availability of foreign savings. Moreover, it is important to be able to defend ITPA-type programs against charges that import liberalization leads to an unacceptably large increase in imports or that reduction of import levies (excluding VAT-type internal taxes) poses an unmanageable threat to the budget. The two approaches developed in Chapter 3 for analyzing the external current account can be very helpful in both regards.

Recall that the first approach defined the current account as the sum of the resource balance, net factor service payments (including remittances), and other net transfers:

(5-1) $\quad CA = [X-M] + FS_n + FT_n$, where
$\quad X = $ exports
$\quad M = $ imports
$\quad FS_n = $ net factor services
$\quad FT_n = $ other transfers, net.

The second approach defined the current account as the sum of net financial savings by the government and nongovernment sectors (including public enterprises).

(5-2) $\quad CA = [S_g - I_g] + [S_{ng} - I_{ng}]$
(5-3) $\quad CA = [T - G] + [S_{ng} - I_{ng}]$
where
$\quad S_g = $ government (budgetary) savings
$\quad I_g = $ government investment
$\quad T = $ government receipts
$\quad G = $ government expenditures
$\quad S_{ng} = $ nongovernment savings, including by public enterprises
$\quad I_{ng} = $ nongovernment investment.

Evaluation of the Current Account: The First Approach.

Between 1984 and 1987, net factor services—including remittances (FS_n), net transfers (FT_n), and the resource balance (X-M)—all showed marked improvement (see Figure 2.1b). As a result, in 1987, the current account showed a surplus for the first time since 1974, equivalent to about 1.4 percent of GDP.

Regarding net factor service payments, good performance in earnings (essentially worker remittances) was initially matched in 1984 and 1985 by higher outlays, mostly on account of interest payments on public external debt, which continued to rise despite some rescheduling (see Figure 5.1 and Annex 2, Table 14).

In 1986, the stagnation of remittances relative to GDP was more than compensated by the significant rise in rescheduling of interest. As a result, net factor services jumped to 5 percent of GDP as against 2.5 percent the year before. This tendency was maintained in 1987.

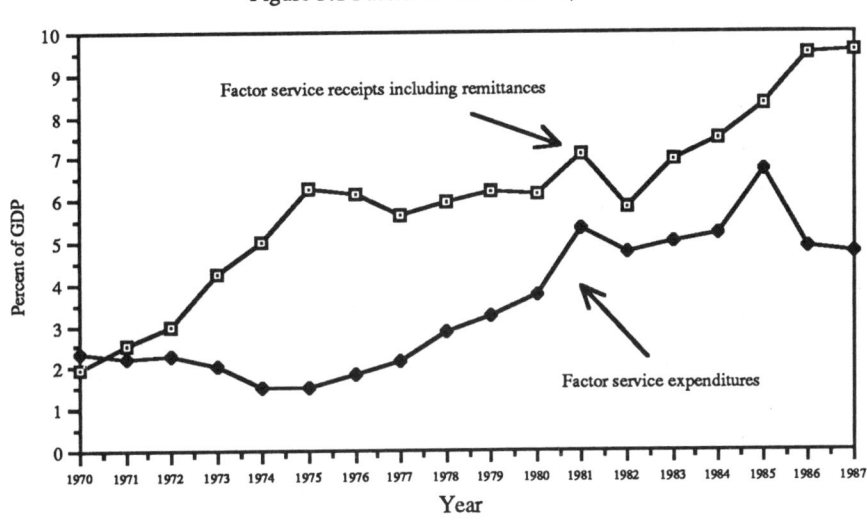

Figure 5.1 Factor Service Balance, 1970-87

The behavior of exports of goods and nonfactor services is summarized in Figure 5.2 (Annex 2, Table 15). Merchandise exports have continued to show variable performance, whereas earnings from nonfactor services have shown a secular rise relative to GDP, essentially because of tourism's excellent performance.

The more detailed information of Figure 5.3 (Annex 2, Table 16) shows once again the sensitivity of overall merchandise exports to the fluctuations of the world market for phosphates and their derivatives. Exports of agricultural and fishing products (including fresh, frozen, and canned fish), as well as nonfood manufactures (excluding phosphates) have grown steadily. Exports of nontraditional manufactures increased by 55 percent between 1982 and 1986; the average annual rate of growth was 15 percent compared with only 5 percent in the preceding four-year period. These trends continued in 1987. Since 1984, Morocco has begun to regain market share in the EC. Moreover, for some products (for example, trousers, skirts, and blouses), duty free quotas to some member countries have been exceeded. This has led, on occasion, to the institution of voluntary export restraints.

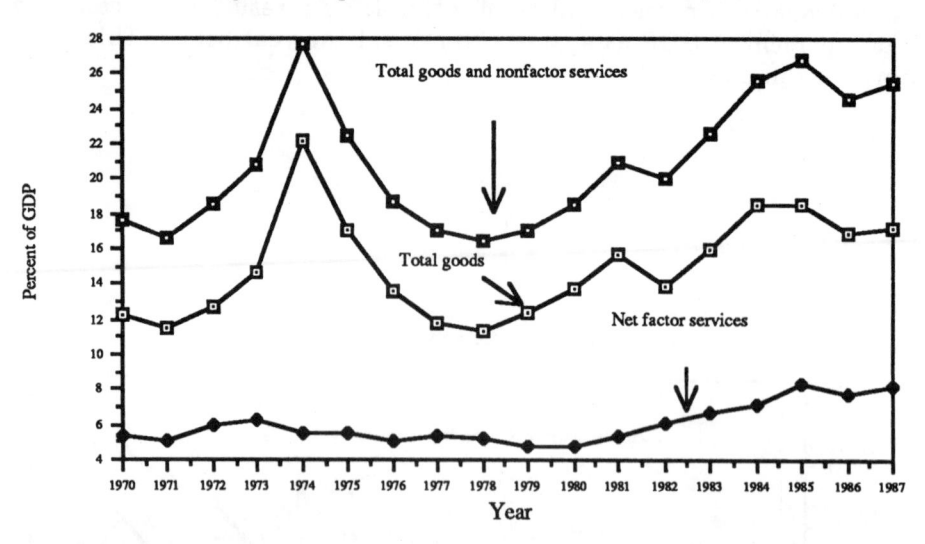

Figure 5.2 Exports of Goods and Nonfactor Services, 1970-87

Export performance has, in fact, been even better than indicated by the preceding discussion because of the important contribution of exports realized on a subcontracting basis, for which payment is received only for the value added to raw materials sent by foreign clients to Morocco for processing. Subcontracting is an important export activity in its own right. The corresponding inflows of

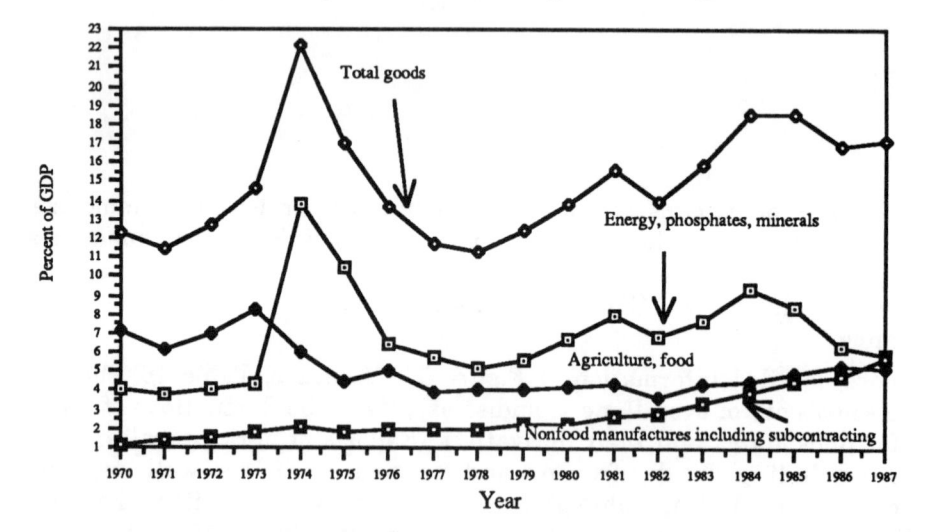

Figure 5.3 Exports of Goods, Including Subcontracting, 1970-87

currency (which, of course, are net foreign exchange earnings) are reported under the heading of services in the balance of payments, and they do not appear in the merchandise statistics. Earnings have more than quadrupled in domestic market prices and have approximately tripled in real terms. In certain activities (for example, clothing and knitwear), subcontracting exports account for an additional

60 percent by weight over and above that recorded in the merchandise trade statistics.[19] In other areas, such as electronics, most exports occur only on a subcontracting basis.

Another striking feature of nonfood manufactured exports excluding phosphates has been their changing composition. In 1987, they accounted for 31 percent of recorded merchandise exports compared with only 20 percent in 1983. The change would be even greater if subcontracting exports were included. During the same period, the share of phosphates fell from 44 to 32 percent. Thus, it is arguable that the volatility or variance of export earnings has been substantially reduced during the ITPA period. Nonfood manufactures including subcontracting are now probably the most important source of foreign exchange earnings.

The strong export performance is attributable to many factors, including active exchange rate management, in line with the hypothesis advanced by ITPA. One recent estimate suggests that the short-run and long-run elasticities are 0.8 and 1.6, respectively (World Bank 1988a).

Total merchandise and nonfactor service imports grew strongly in 1984, leveled off in 1985, and then declined in 1986 and 1987. Government imports—essentially military hardware—continued the downward trend begun in 1982, until 1986, when there was a sharp one-year upturn (see Figure 5.4 and Annex 2, Table 17).

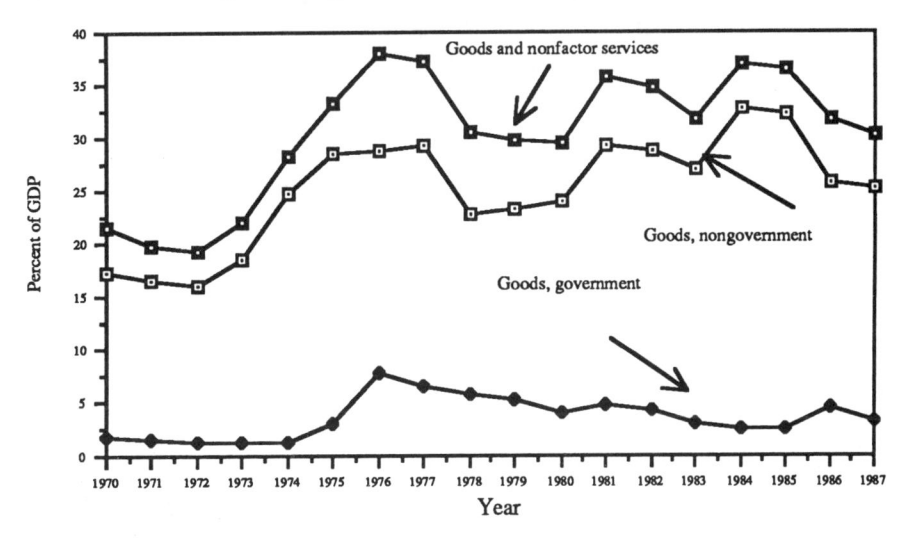

Figure 5.4 Imports of Goods and Nonfactor Services, 1970-87

Some attributed the initial import boom to the ITPA program, which, they said, was causing imports to rise whereas a contraction was needed. These arguments

---

19. The corresponding volume and FOB value of certain categories of merchandise exports, especially garments and knitwear, are therefore significantly underestimated. Consequently, the total FOB value of garments may be between 1 and 1.5 billion dirham more than the recorded value of 2.5 billion dirham.

were incorrect, as may be seen from the detailed decomposition presented in **Figure 5.5** (Annex 2, Table 18) and Table 5.7.

In the first place, some expansion of imports in 1984 was probably inevitable given the highly constrained level of the year before and the corresponding need to replenish stocks. Other major explanatory factors included the resurgence of drought in 1983 that led to higher food imports, especially cereals and sugar. Capital goods increased as the result of the public investment program. The 22 percent rise in the import to GDP ratio for intermediate goods from 8.6 to 10.5 percent a year later reflected increases of 36 percent for edible oils and 66 percent for sulfur compared with only 18 percent in other categories. Imports of energy increased by 15.5 percent and imports of consumption goods by 7.5 percent.

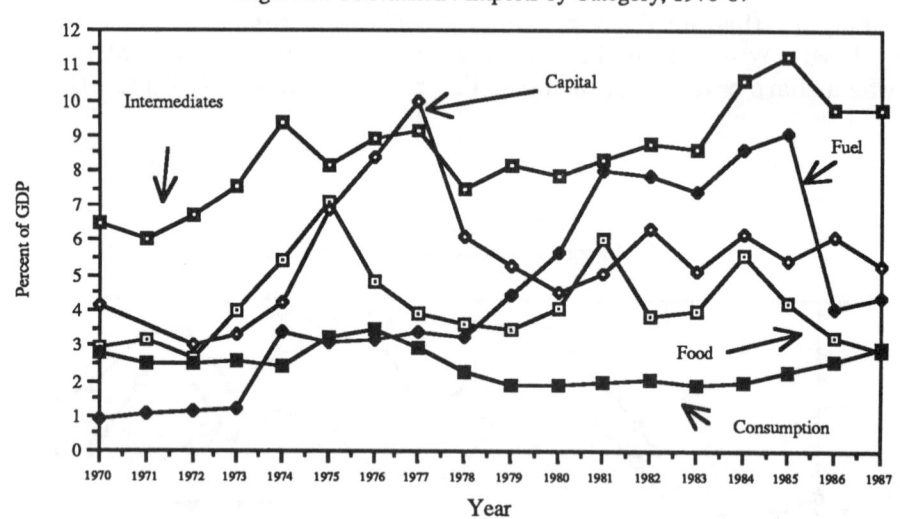

Figure 5.5 Merchandise Imports by Category, 1970-87

Table 5.7 Import to GDP Ratio, by Import Regime, 1983-87

Import regime	1983	1984	1985	1986	1987
Temporary admission	2.5	3.0	4.0	3.9	4.3
Domestic market	24.5	29.9	28.3	21.9	20.9
Total	*27.0*	*32.9*	*32.3*	*25.8*	*25.2*

In 1985, import demand slackened for food and capital goods but increased for consumption and intermediate goods (on account of a continuation of higher imports of edible oils and sulfur). In 1986, the major factors were the collapse of petroleum prices and a further decline in food imports attendant upon a bumper harvest. Expenditures on edible oils also declined along with international prices. These tendencies continued in 1987. But consumption goods showed a steady rise.

Explanatory factors for the evolution of capital intermediate and consumer goods are as follows. The sharp advance in capital goods in 1984 resulted from orders placed by OCP and other enterprises, such as the railways and the telephone company, independently of and before the inception of ITPA. These advances tapered off during 1985. At the same time, imports of industrial capital goods realized under the auspices of the investment code declined sharply relative to GDP, attendant upon a sharp shift in incentives from capital to labor.[20] Among intermediate goods, edible oils and sulfur showed considerable growth in 1984 and 1985 on account of price explosions in international markets. But imports of these goods would have increased regardless of ITPA. Edible oils are considered essential to consumers and are imported as needed regardless of cost. The same is true of cereals and sugar, and of sulfur, which is indispensable in the production of phosphoric acid.[21]

Imports for consumption goods were also subject to the same influences as some intermediate goods that are incorporated into Morocco's manufactured exports: they rose along with exports rather than because of import liberalization.

Import behavior should not be measured by the growth of total imports but rather net of those that are reexported after processing. Imports so adjusted give a true representation of the evolution of imports for home consumption. As Table 5.7 shows, these declined after 1984.

OVERALL COMMENTS AND IMPLICATIONS. The 1983-87 period was marked by favorable developments regarding exports and imports of goods and services. This can be seen from a fairly detailed examination of the principal components of inflows and outflows of foreign exchange, rather than from a review of the performance of simple aggregates.

On the export side, the growth and diversification of nonfood manufactures excluding phosphate derivatives were significant factors in reducing the instability (or variance) of foreign exchange earnings. Previously, these earnings had been dominated by phosphates and agricultural commodities. Both were highly susceptible to price or climatic fluctuations or both.

In a similar vein, the importance of closely examining the performance of nonfactor service exports, especially tourism and subcontracting, also emerged. The growth of the latter has been spectacular since 1983. Subcontracting, which is, in fact, manufacturing, consists of processing raw materials sent to Morocco,

---

20. In part, this corresponded to the reform of the investment code in 1983, which had reduced incentives to capital and increased those in favor of labor.

21. Imports of these commodities and total imports were as follows in 1983-87 billions of dirham:

	1983	*1984*	*1985*	*1986*	*1987*
Sulfur	1.0	1.4	2.3	2.2	2.1
Edible oils	0.6	1.1	1.4	1.0	0.6
Cereals	2.0	3.6	2.8	1.7	1.6
Total	*3.6*	*6.1*	*6.5*	*4.9*	*4.3*
Total imports	*27.0*	*32.9*	*32.3*	*28.3*	*25.2*

generally by European and American importers. Since the exporter receives payment only for the value added corresponding to his processing costs, the foreign exchange earnings are recorded as a service. The corresponding tonnages are not recorded in merchandise exports. As a result, the official trade statistics substantially understate actual export volumes. This kind of activity, which involves the marketing of a service rather than products, can be an important source of foreign exchange earnings. Some firms reported that subcontracting to several different clients was more profitable than manufacturing and marketing goods. Consequently, trade policy should ensure that production and export incentives do not discriminate against subcontracting.

On the foreign exchange expenditure side, detailed analysis is also important since each subcategory of import may respond to many different factors, of which import liberalization is only one. In particular, it is essential to distinguish between total imports and imports for use in the domestic market. The former may rise and the latter fall, even during a liberalization program, on account of increasing reexports of some imported goods that are processed into exported manufactures. In the Moroccan context, this effect actually occurred for some intermediate and consumer goods. Capital goods imports also rose, in part because of OCP's massive investments to increase exports of phosphoric acid and in part because of the public investment program that was determined politically rather than by market forces. In other cases, fluctuations in import to GDP ratios corresponded in some degree to variations in world prices (edible oils, sulfur, crude oil, sugar, and wheat). Domestic price policy can also have a major impact on imports of consumer goods with regulated prices, even if imports are subject to licensing. This was the case for edible oils; prices were not adjusted upward in line with world prices. Had prices been allowed to rise, foreign expenditures on this commodity would necessarily have grown more slowly and perhaps even have declined. Similar considerations apply to wheat and sugar, where domestic prices are insulated from world prices and all demand is met as a matter of public policy.

Finally, the overall evolution of imports depends in part on budgetary expenditures, including on foreign exchange, which may be assumed to be politically determined outside of the import licensing regime. The influence of the government budget on the external current account is addressed in the next section.

Evaluation of the Current Account: The Second Approach

One of the major criticisms of the ITPA program was that it had a negative impact on tax receipts and hence the budget. This contention may be conveniently examined by decomposing the external current account gap into the sum of net nongovernmental financial savings—that is, nongovernment savings less investment ($S_{ng} - I_{ng}$)— and the budget surplus or deficit (T - G). This approiach can also be used to highlight the individual behavior of budgetary and nonbudgetary savings and investment. The starting point for the discussion is Figure 2.1a. It showed, in broad terms, that both net nongovernmental financed savings and the budget deficit improved during the life of the ITPA. Domestic and national savings improved from 11.6 and 14.3 percent respectively in 1983 to 14.1 and 18.8 percent in 1987. More specifically, the gap between nongovernment savings and investment showed a sharp improvement in 1984 and thereafter a steady upward trend. The fluctuations in the private (that is, nongovernment) saving-investment gap may be attributed to fluctuations in nongovernment investment. This showed a sharp rise in 1984-85, as a result of OCP's investment in new phosphoric acid plants, before declining. Other productive sector investment actually fell as a percentage of GDP during the whole period.

The overall Treasury deficit, after rescheduling, declined from 11.1 to 5.4 percent of GDP between 1983 and 1987. The Ministry of Finance argued that there could have been a greater reduction but for ITPA's reduction of the SIT.

During this 1983-87 period, expenditures on goods and services before amortization fell substantially relative to GDP as a result of the government's stabilization efforts. But gross demand including amortization did not show a commensurate decline (see Figures 2.8 and 2.9). Inadequate domestic financial savings, net of investment, forced foreign creditors to consent to substantial rescheduling of interest and principal (about 10 percent of GDP) in 1986 (see Figure 2.8, shaded area).

The primary means of government adjustment was the reduction of investment expenditures, which fell to only 5 percent of GDP in 1986, rather less than in 1973.[22] Yet recurrent expenditures excluding interest were over 18 percent of GDP in 1987, still approximately 25 percent more than in 1973. Consumption subsidies rose in terms of GDP in 1984 and 1985, whereas under the IMF stabilization program they were supposed to decline.

Government receipts relative to GDP showed relative stability (22.4 percent) until 1986, when there was a decline of 1.2 percent. As indicated earlier, this decline was attributed solely to ITPA's reduction of the SIT. However, 1986 was a year in which the SIT remained unchanged. Hence, an alternative explanation must be sought. In the first place, 1986 like 1985 was a year of strong growth (+12.6 percent), particularly in agriculture and nonphosphate manufactured exports that are virtually exempt from direct taxes. At the same time, OCP transfers to the budget were virtually eliminated because of depressed phosphate markets. Receipts from internal indirect taxes stagnated on account of difficulties encountered in transforming the TPS into the VAT. Finally, imports declined by 13.6 percent in the wake of the stabilization program and the slump in international oil prices. To compensate, the government introduced a special levy on imports to capture the difference between the international price for crude oil and domestic selling prices that were maintained unchanged.

The adverse fiscal impact of the SIT's phased reduction can be partially ascribed to the failure to institute offsetting increases in the TPS and the effective rate of taxation applied to income derived from exports and the investment code corporate income, as had been advocated during the preparation of the ITPA program. This preparatory work had indicated that there were major distortions in import duties, as well as in indirect and direct taxes, which adversely affected resource allocation and tax collections.

The main findings, as updated by subsequent work, were that all tax schedules, whether direct or indirect, combined too many exemptions with unduly high rates on nonexempted income or transactions. In the case of indirect taxes, this was compounded by the existence of consumption subsidies that led to negative rates of taxation for certain commodities. Finally, collections depend too much on imports and not enough on domestically generated value added. This clearly poses a problem in an economy in which imports are called upon to contract relative to GDP if stabilization and adjustment are to succeed.

---

22. Capital expenditures, as resported by Treasury statistics, include transfer to public enterprises for capital formation.

Figure 5.6 (Annex 2, Table 19) presents a decomposition of taxes for the period 1972 to 1987, including the special petroleum levy from 1986 on. The revenue dominance of indirect taxes in general and import levies in particular stands out. Except for 1974 and 1975, direct taxes (D_1) averaged between 4 and 5 percent of GDP and public enterprise dividends (D_2) averaged between 2 and 3 percent. Indeed, public enterprise dividends fell to less than 1 percent of GDP in 1986. At the same time, internal indirect taxes (I) and import levies (M) had much higher ratios that rose relative to GDP pari passu with the import expansion of the 1970s and early 1980s. After 1983, ratios declined with the advent of stabilization and adjustment, which led to declining imports, relative to GDP, and so lower import duty collections.

Collections of direct taxes including dividends have been strongly influenced by OCP transfers to the Treasury. These were considerable in 1974, 1975, 1981, 1984,

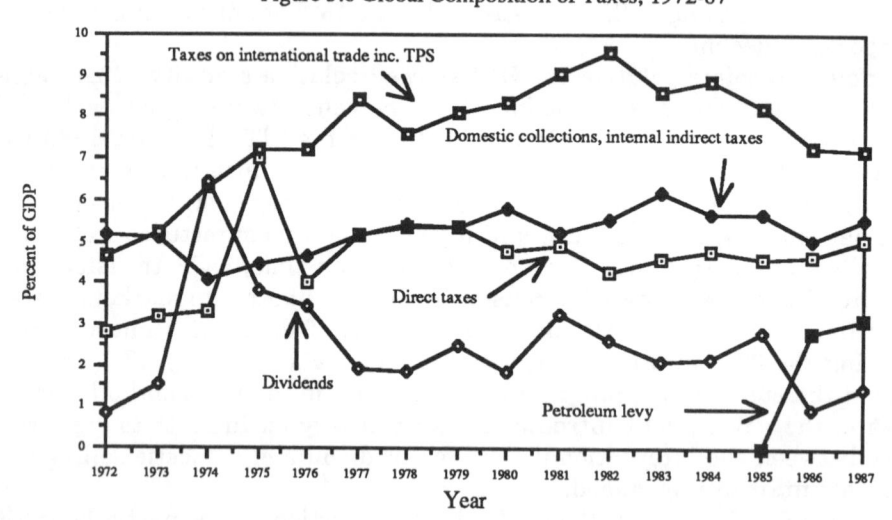

Figure 5.6 Global Composition of Taxes, 1972-87

and 1985—boom years in world markets for rock phosphate or phosphoric acid or both. In 1984 and 1985, OCP transfers amounted to 1 and 1.5 percent of GDP respectively. However, in 1986, the situation deteriorated sharply: for the first time since 1972, OCP made no budgetary transfers at all, as OCP was squeezed between declining output prices and sharply rising world prices for sulfur, one of the primary inputs in the manufacture of phosphoric acid.[23] An additional factor may have been outlays related to the construction of a new phosphoric acid plant. But in 1987, OCP again made positive transfers (albeit less than 30 percent of the 1985 level). Apart from OCP, the only other public enterprises to have made regular and significant budgetary transfers were Bank Al Maghrib and, to a lesser extent, one of the state-owned commercial banks. In the case of Bank Al Maghrib, no transfer occurred in 1986.

---

23. Between 1984 and 1986, foreign exchange earnings from exports of rock phosphate declined from $925m to $750m, while outlays on sulfur rose from $165m to $250m. This resulted in a $260m decline in net foreign exchange earnings (about 1.7 percent of GDP.

During preparation of the ITPA programs, considerable attention was devoted to trying to ensure satisfactory growth of direct tax revenues, especially from export-generated profits. The problem was that the 1973 export code exempted all export-related profits from the 48 percent profits tax. The total direct tax burden on export-related profits was therefore only 4.8 percent on account of the Participation à la Solidarité Nationale (PSN) as compared with 52.8 percent (PSN included) for domestically generated profits. To redress this situation, which was important given ITPA's export orientation, it had been proposed by the Ministry of Commerce and Industry to significantly lower the standard rate of profits tax and to raise the PSN by an equivalent amount. The idea was that, for domestically generated profits, the combined incidence of the two would remain at 52.8 percent, whereas for export-related income it would rise pari passu with the PSN.[24] In this way the effective rate of profits taxation could easily have been increased to, say, 20 percent without the government having to tackle the politically difficult task of abrogating the export code. However, this proposition was not retained by the Ministry of Finance despite repeated insistence on its potential as a revenue-raising device. A similar proposition was adopted in the 1989 budget law more than five years later.

Figure 5.6 shows that import levies (M)—including TPS and internal indirect taxes (I)—have declined relative to GDP since 1984. A decline in import levies would have been expected in any case, since imports themselves contracted from 29.8 to 21.2 percent of GDP under the impetus of the stabilization program and the decline in crude oil prices. The decline was accentuated, of course, by the reduction of the SIT and import duties. It was expected to be offset by an increase in TPS rates with the transformation of the TPS into the VAT. But this transformation, a building block of the adjustment program, did not materialize to the dismay of the pro-ITPA forces.

Collections from excises and the TPS, including on imports, also declined relative to GDP. In the case of excises, the decline was due to the specific nature of these taxes and the failure to augment them in line with inflation. This failure had been especially important for petroleum and sugar. The situation was partially rectified in 1986 by the introduction of the special levy on petroleum following the drop in the international price of crude oil. The revenue yield was equivalent to a little less than 3 percent of GDP. No changes were introduced for sugar, however.

The decline in domestic TPS receipts relative to GDP was initially caused by unusually large payments to exporters, and to OCP in particular, of restitutions claimed for the preceding year (1984).[25] In 1985 and 1986, factors contributing to the decline in TPS receipts were (a) the growth, relative to GDP, of exports and agriculture, both of which were statutorily exempt from TPS; (b) the stabilization and then the absolute decline in imports attendant upon stabilization and decline of the world price of crude oil; (c) the introduction in 1986 of the value added tax that

---

24. Note that this would also have increased revenues from new firms serving the domestic market since *all* new firms were exempted from the profits tax for two years under the investment code. But this exemption did not apply to the PSN.

25. Note, in this connection, that TPS receipts consist of the sum of gross collections on imports and *net* collections on domestic transactions. This is because domestic collections equal the difference between receipts on sales and deductions, including restitutions, on purchases. Net collections then depend on the speed with which deductions are allowed, or restitutions are paid. Similarly, net domestic collections may be adversely affected by arrears if the seller remits the related tax to the Treasury when he is paid, rather than on delivery regardless of the terms of payments he grants to the purchaser.

replaced the TPS both on imports and domestic transactions; and (d) the government arrears problem.

The VAT reform generalized the right of producers to deduct the tax paid on purchases from that levied on sales, and it reduced the waiting time before deductions could be credited. Both led to a decline in net collections for the government and should have been compensated by an increase in rates. Such an increase was also needed to compensate for the reduction in the SIT. But, in fact, rates may have been reduced rather than increased on average, and the number of exemptions was not reduced.[26] It seems to have been believed that the extension of the VAT to wholesalers would provide adequate supplemental resources. But here, too, things went wrong. Wholesalers refused to pay the VAT on their existing stocks as these already incorporated TPS; they argued correctly that payment of VAT under these conditions would amount to double taxation.[27]

Finally, all of the above may have been compounded by the massive government and public enterprise problem that emerged in 1984. This is because the TPS or VAT remittances to the Treasury can be made on a payment basis rather than on an invoice basis (that is, when the seller is paid by his client, rather than on delivery of the goods). Under these conditions, arrears in payment of enterprises automatically generate arrears in remittances to the Treasury.

No discussion of developments from 1984 to 1987 would be complete without comparing the relative importance of domestic indirect taxes and consumption subsidies, which are, of course, negative taxes (see Figure 5.7 and Annex 2, Table 20). In 1974, these subsidies exceeded total collections from domestically levied TPS and excises. After reaching a low in 1978, consumption subsidies rose steadily until 1981, when social unrest peaked. The same pattern was evident in 1984 and 1985 following the resurgence of unrest in January 1984. In 1981 and again in 1984, outlays on subsidies exceeded net domestic collections of TPS. In 1986, consumption subsidies declined to about 1.5 percent of GDP. There was a further reduction of 1 percentage point in 1987. But consumption subsidies continue to represent a substantial drag on the budget.

OVERALL COMMENTS AND IMPLICATIONS. This approach to analyzing the current account deficit has revealed the complexity of the forces that influenced its evolution during the period 1984 to 1987. In particular, it is evident that the reduction in the ratio of tax receipts to GDP cannot be attributed simply to the reduction of the SIT. It would be more accurate to say that no serious attempts were made to use the tax reform to raise internal indirect taxes by replacing the TPS with higher VAT rates. This would have continued the successful precedents of January 1982 and July 1983, when the Ministry of Finance was persuaded that it was better, both on revenue and economic grounds, to raise the TPS rather than the SIT. The real reason why this approach was not maintainesd lies in the politics, internal and external, of the tax reform. The outcome might have been different if the ITPA

---

26. Under the TPS, butter, tea, coffee, and spices had been taxed at 12 percent. The VAT taxed the last three items at 7 percent and exempted butter. Exemptions for flour, milk, sugar, etc. were maintained. On the other hand, transport now pays 7 percent as against 4 percent under the TPS. Notwithstanding these changes, domestic VAT collections were essentially unchanged in 1986 compared with 1985. In 1987, they rose by 3.3 percent. By contrast, in 1982, when the standard rate of TPS was raised from 15 to 17 percent, domestic collection rose by 31 percent and import collections by 38 percent.

27. Consequently, an exceptional credit had to be introduced for the TPS already paid on stocks.

protagonists had been able to exercise a greater influence over the preparation of the tax reform that was managed, internally and externally alike, without enough concern for the unique opportunity it represented to improve both the structure of receipts and resource allocation.

This is an important lesson. Trade policy reforms that aim to reduce import levies must be simultaneously accompanied by substantial adjustments in internal taxes, both direct and indirect, so as to increase the share of receipts derived from taxes on (a) profits and wages and (b) final consumption. In connection with the latter, reduction in consumption subsidies should also have a high priority. The ITPA experience suggests that these adjustments, or tax reforms if necessary, cannot be deferred to a later phase of the adjustment program. To do so basically undermines the durability of the reduction in protective import duties that is so essential to ameliorating competitiveness and increasing the efficiency of

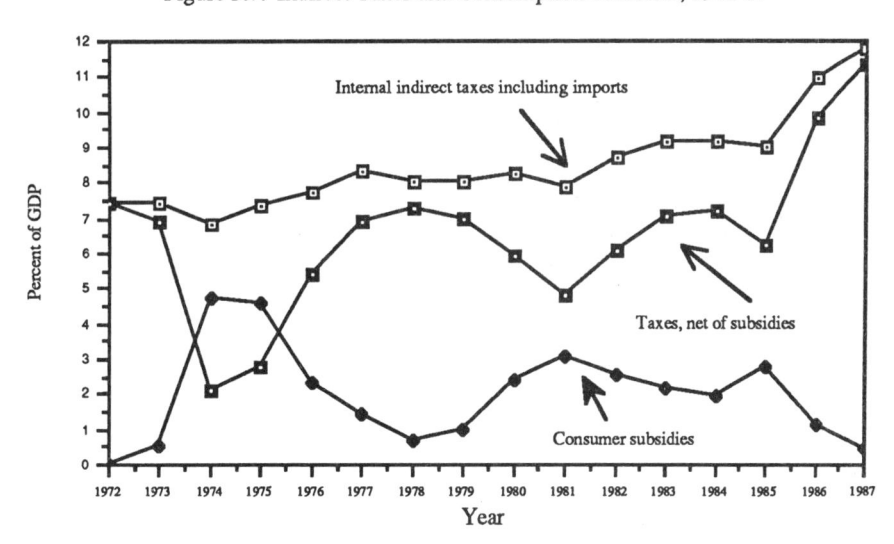

Figure 5.7. Indirect Taxes and Consumption Subsidies, 1972-87

resource allocation in import substituting sectors. In Morocco, the conditions for success were met because of careful technical preparation before the ITPA program. Unfortunately, the political will to continue this process during the ITPA program evaporated.

Microeconomic Consequences of ITPA

Did ITPA improve resource allocation, employment, factor productivity, and Morocco's economic and business environment? An exhaustive answer to this question is not possible in this study, although it has been attempted in a recent evaluation of the impact of liberalization on trade and industrial sector adjustment (World Bank, 1988a). That report concludes that real wages fell during the course of adjustment, but this was in part compensated by employment growth, particularly in labor-intensive, export-oriented activities.[28] The results of the analysis are inconclusive concerning ITPA's impact on factor productivity, which perhaps is not surprising in view of its recent adoption.

More selective evidence is available from a sample of exporting and import substituting firms. For example, there is universal agreement that the administrative cost of exporting and importing has substantially declined with the simplification of international trade procedures. Port clearance times have been reduced by 50 percent to six days for export-related industrial imports. Some exporters report that access to working capital has improved. The reform of export insurance is, as yet, incomplete and continues to pose problems for exporters.

The liberalization of competing imports has substantially reduced profit margins and induced firms to try to cut costs. To some extent, this process has been impeded by restrictive regulations concerning labor force reduction. Liberalization has completely eliminated costly domestic content requirements in the paper industry. Domestic manufacturers are now free to choose their raw materials, domestic or imported, as they wish. Yet domestic content requirements continue to exist for industries that are still protected, such as steel tubes and pipes and automobiles, with the result that many domestic products are often not made using international quality norms. On the other hand, the elimination of administrative controls on imports effected under the temporary admission customs regime has greatly reduced the costs of some raw materials used by exporters who may now import at world market prices if they wish.

As import restrictions have eased, other problems have come to the fore—barriers to entry and exit, bankruptcy provisions, domestic transportation, access to working capital credit, availability and cost of industrial land and related infrastructure, the cost of electricity, and the quality and cost of labor. Similarly, there is now a resurgence of debate on the desirability of continuing to restrict foreign investment through the 1973 Moroccanization Law.

One of the most interesting aspects of the ITPA process has been to see how cognizance of these problems has grown with the implementation of the ITPA program. Different actors in the Moroccan economy have come to realize that these problems, perennial in nature, as well as their impact on the economy, have been masked by the protection issue. Some have even stated that slowing down liberalization would be undesirable because it would enable these problems to slip from the public view, whereas they need to be addressed urgently.

Conclusion

This chapter has examined in detail IMF and World Bank approaches to stabilization and adjustment from 1983 to 1987. The approaches of both institutions have evolved in line with their increasing understanding of the complexity of Morocco's economic problems.

Discovery of an important arrears problem, both domestic and international, prompted the IMF to adjust its financial programming approach. Specific targets were introduced into post-1983 standbys for internal and external arrears reduction as well for Treasury net borrowing including arrears repayment. Essentially, quarterly ceilings were established in domestic currency for the

---

28. The real devaluation of the dirham on purchasing power was mitigated because consumer prices rose more slowly than did producer prices. The "internal" real exchange rate (that is, the relative price of traded to nontraded goods) tended to depreciate.

Treasury deficit on a cash basis including repayment of arrears. Conditionality was consequently much stricter than in the pre-1983 period, when the performance criteria related only to ceilings on bank credit to the economy as a whole with a subceiling for the Treasury.

After 1983, the World Bank's approach also became much more focused. The ITPA program concentrated on in-depth treatment of two themes: (1) price and nonprice incentives pertaining to production in general and exports in particular and (2) interest rate levels and structure, as they affect savings and investment, both in real and financial terms. The primary macroeconomic objective was to increase the domestic supply of foreign exchange from economically efficient exports and import substitution. The means to this end was a comprehensive package of export promotion measures undertaken with a phased import liberalization program and a reform of tariffs and indirect taxes. Specifically, the tariff reform aimed at a gradual reduction of the maximum rate of duty to 25 percent.

The Bank carefully analyzed the public finance consequences of the proposed adjustment program. All proposals to reduce import duties and their surrogates were accompanied by recommendations to raise internal taxes and cut expenditures. However, this advice was disregarded on the grounds that it was politically impossible to increase such sensitive taxes as the TPS and the PSN. But, in both cases, the historical precedents of the 1980-82 period were overlooked—when the TPS was raised on two occasions and the PSN created.

The result was that ITPA's impact on the budget deficit may been worse than was originally anticipated. However, the failure to raise the TPS and PSN implied that ITPA's supply-side program was also rather stronger than originally envisaged. In other words, output growth may also have been stronger than foreseen.

Another significant dimension of the post-1983 approach to adjustment concerns the importance of Bank/Fund cooperation, particularly concerning the public finance and arrears problems. In general, cooperation was good, but sometimes there were differences in priorities. For example, there were protracted discussions, ultimately unsuccessful, as to how to recover the shortfall resulting from the reduction of the SIT and import duties. On the arrears issue there was greater success, with the IMF concentrating on the identification and management of arrears. In turn, the Bank in ITPA2, and especially in the subsequent Public Enterprise Rationalization Loan (PERL), embarked on ways (a) to prevent the growth of arrears particularly as regards the investment budget and (b) to keep arrears reduction from having adverse macroeconomic effects that could result from large amounts of liquidity being injected into the economy. It may be added that PERL set out to examine in great detail the microeconomic problems of public sector enterprises—an indispensable component of the adjustment process in Morocco.

ITPA was an approach to adjustment based upon collaborative analysis and research by the Bank and the Government of Morocco over an extended period from 1979 to 1983. During this period, Morocco's industrial sector problems were analyzed in depth by a joint Government/Bank research team based in the Ministry of Commerce and Industry. As a result, the Bank had a day-to-day view of problems as perceived by Moroccan civil servants, industrialists, and consumers. By the same token, Moroccan authorities felt that the adjustment program, even if it did not meet with universal support, had been based on unusually extensive, joint analysis of the country's underlying problems in the sectors covered. As a result, it was considered that the proposed adjustment program had been largely designed in Morocco. This was how the program was presented to the industrial sector, and it undoubtedly helped in the generation of the consensus necessary to stay the course during a rather difficult period.

6
Lessons from the Case Study

The preceding two chapters focused on the Moroccan government's approach to stabilization and adjustment in the 1973-87 period, as they were influenced by the International Monetary Fund and the World Bank. The purpose of this chapter is to compare the approaches of the Fund and the Bank, in order to highlight the lessons that may be useful to policymakers in other countries.

The IMF's Approach to Stabilization and Adjustment

The IMF initially emphasized that growth would be compromised unless domestic savings, especially budgetary savings, were increased through higher taxes and lower expenditures. These measures were to be accompanied by tight monetary policy to restrain private demand and imports. Consequently, there was little opposition to the Moroccan government's strategy of raising trade taxes and tightening import controls between 1975 and 1979, despite the negative supply-side impact of these actions. However, by the end of 1979, it became apparent that a policy of budgetary stabilization and expenditure reduction was not sufficient to improve Morocco's deteriorating economy and that a supply-side program was needed to increase output. This recognition led to the 1980 Extended Fund Facility (EFF) and made the International Monetary Fund a much more active participant in the discussion of Morocco's economic and budgetary policy.

Yet, as indicated in Chapter 4, the EFF contained hardly any supply-side measures. Rather, the program emphasized the need (a) to continue to contain public and private consumption, and (b) to increase investment in import substitution and energy, even though these measures might not bear fruit until after the end of the program period. The EFF made scant effort to reduce the anti-export bias that had resulted from the exchange rate appreciation and increased tariff and nontariff import restrictions of the preceding decade. Indeed, there was relative pessimism about export growth on the part of the Moroccans as well as the Fund.

In the preparation of the program, extensive use was made of the Fund's financial programming model. Performance criteria included

- quarterly ceilings on total domestic credit from the banking system with a subceiling for the Treasury;

- a ceiling on new external loans from international financial markets with maturities between one and ten years;
- understandings as to the future conduct of exchange rate policy and target levels for the above-mentioned performance criteria.

In addition, it was expected that a comprehensive reform of direct and indirect taxes would be submitted to Parliament during the life of the program.

The EFF became effective in late 1980, but was canceled a year later because the Moroccan government's domestic and foreign borrowing exceeded the agreed ceilings by 3 and 50 percent, respectively. These overruns occurred only partly because of a poor harvest and an associated jump in imports. The primary reason was the government's inability to halt the continued rise in capital and recurrent expenditures (Figures 2.8 and 2.9). Attempts to reduce consumption subsidies and transfer payments via sudden, large price increases led to substantial consumer unrest.

The EFF was canceled in late 1981 and replaced in April 1982 by a more conventional standby that contained more or less the same conditionalities as the EFF. The standby emphasized the need to reduce the budget deficit through lower expenditures and higher tax revenues. The supplemental receipts were obtained, not by raising import levies—in particular, the special import tax (SIT) and stamp duty—as had been done in the past, but by raising internal indirect taxes. The decision to raise the standard rate of the manufacturer's sales tax (TPS) rather than the SIT from 15 to 17 percent was the result of strong pressures from the Ministry of Commerce and Industry. It argued that increases in customs levies would only increase anti-export bias, whereas this needed to be reduced, and that raising the TPS would raise revenues without augmenting anti-export bias. However, current expenditures were cut without any price adjustments on subsidized consumer goods. Some attempt was made to contain transfers to public enterprises. The Moroccan authorities again promised that the tax reform would be submitted to Parliament.

In the event, expenditure control proved exceedingly difficult to enforce, and the targets for the current account and budget deficits were largely exceeded. The performance criteria, especially relating to foreign borrowing, were met only because newly contracted loans were on concessional terms, for which the program contained no ceiling. In addition, external arrears had accumulated for the first time. Nevertheless, the situation appeared to be contained by the end of 1982. Total government expenditure including amortization of debt had leveled at about 42 percent of GDP (see Figure 2.8). The standby was successfully completed.

IMF operations following the 1983 foreign exchange crisis differed significantly from preceding ones, both on the supply and demand sides. Regarding supply, the programs insisted on substantial exchange rate depreciation and its subsequent management on a flexible basis, as well as on the adoption of export promoting measures identified by the Moroccan government and the World Bank during the preparation of the World Bank-financed Industry and Trade Policy Adjustment program (ITPA). Interest rates were increased to stimulate financial savings, especially in the form of worker remittances. Another critically important contribution by the Fund was the mobilization of external resources in the form of debt rescheduling by bilateral and commercial creditors (see Figure 2.8). On the demand side, the IMF programs emphasized, of course, the need to reduce budgetary expenditures particularly on consumption subsidies and transfers to public enterprises, although little attention was given to what might be done to improve the efficiency and productivity of the latter. The budgetary components of the post-1983 standbys had a much sharper focus as the massive accumulation of arrears, particularly on the domestic side, became evident.

Consequently, the standby programs incorporated performance criteria relating either to arrears accumulation or repayment. Coupled with limits on Treasury access to domestic and foreign borrowing, this became equivalent to introducing a performance criterion for the budget deficit on a cash basis after settlement of arrears.[1] As a result, there was a need for even stronger revenue-raising and expenditure-reducing measures. Surprisingly, on the revenue-raising side, the Fund did not push as hard as it might have for the rapid implementation of the tax reform that would have shifted the burden from trade taxes (import duties) to indirect taxes. Inevitably, lack of progress on this front induced growing opposition to reduction of trade taxes, which the Ministry of Commerce and Industry and the World Bank considered essential for reducing anti-export bias and, in the longer run, for establishing neutrality of incentives for import substitution and exports. Moreover, when finally adopted in 1986, the VAT reform did not contain an upward adjustment of TPS rates even though this was clearly justified and probably feasible politically given the successful increases in the TPS in 1982 and 1983.

As a result, the government came under increasing pressure in 1986 and 1987 to arrest and then reverse the decline in trade taxes that had begun in 1984 as part of its industry and trade policy reforms.[2] At the same time, the Fund continued to push for cuts in consumption subsidies, transfers to public enterprises, and capital expenditures. Reducing consumption subsidies may reduce distortions in the tax structure to the extent that an untargeted consumption subsidy is considered a negative tax. It did not prove easy to prune capital expenditure in part because the government did not dispose of adequate controls over the expenditure process. Once committed, expenditures proved exceedingly difficult to stop, thus leading to an accumulation of arrears when resources were not available for payment. An important contribution of the IMF program was to identify this problem and to insist on the implementation of an effective solution via the adoption of improved budgetary control procedures.

How can the Fund's approach be described in terms of the methodology, presented in Chapter 3, for analyzing stabilization and adjustment programs? The starting point for this methodology is the decomposition of the current account as follows:

$$\begin{aligned} CA &= (X - M) + FS_n + FT_n \\ &= (S_g - I_g) + (S_{ng} - I_{ng}) \\ &= GNP - E \end{aligned}$$

where X denotes exports; M, imports; FS_n, net factor service payments; FT_n, net foreign transfers; S_g, government (budgetary) savings; I_g, government investment; S_{ng}, nongovernment savings; I_{ng}, nongovernment investment; GNP, gross national product; and E, national expenditure on goods and services.

The preceding discussion suggests that the Fund's emphasis has been on reducing the current account gap through compression of the budget deficit $(S_g - I_g)$.

---

1. Prior to 1983, the programs specified a target for the budget deficit as a percent of GDP.
2. An increase in trade taxes occurred in January 1988, when the SIT, levied at 5 percent, was replaced by a new tax—"prélèvement fiscal à l'importation" (PFI)—of 12.5 percent.

The means to this end have been to raise taxes and lower current expenditures so as to raise government savings (S_g). The Fund has also pushed, perhaps reluctantly, for cuts in investment expenditures (I_g), particularly when the level of expenditure had been leading to arrears accumulation because of a shortage of financial resources. Fund programs initially concentrated rather less on how to raise nongovernmental savings (S_{ng}); it was hoped that savings would be achieved through a restrictive monetary stance and by raising interest rates. Only after 1983 was an active exchange rate policy strongly encouraged as a means of raising the growth rate of tradables, and hence nongovernmental savings, through higher exports of all goods and services and through worker remittances.

In retrospect, the importance of this emphasis on budgetary stabilization is clear. Reducing the external current account deficit quickly to manageable proportions would have been impossible in the absence of a drastic reduction in the budget deficit, as is apparent from Figure 2.1a. The concentration on the domestic arrears problem was equally important in 1984 and 1985, when budgetary and public enterprises arrears, both between themselves and the rest of the economy, endangered the economic health of productive sectors.

In general, the Fund's approach remained essentially macroeconomic although its programs recently have become more detailed, particularly regarding budgetary policy. The target for the budget deficit that includes arrears accumulation and settlement is particularly useful since it highlights total government demand for resources. In turn, reliable information about total budgetary needs is essential for the design of structural reforms in nongovernmental (that is, productive) sectors that aim, through exchange depreciation and lower import levies, to reduce bias against production of tradables.

In Fund programs it is generally left to government authorities to design the specific measures needed to ensure compliance with performance criteria, particularly in the budgetary area. For example, the Fund often appears relatively indifferent about how tax revenues are raised, even if the measures may have significantly adverse effects on the efficiency of resource allocation, as in the case of trade taxes. Little attention also is given to whether public enterprises will adjust to a reduction in budgetary transfers and if so how. These are important omissions because improvement in budgetary savings, and hence in the budget deficit, attained through reduction in transfers to consumers and productive sector enterprises will have an equivalent beneficial impact on the external current account gap only if it does not cause an equal reduction in nongovernmental savings. In turn, avoiding this pitfall would require an improvement in the efficiency of resource use in nongovernmental sectors. Consequently, greater attention could usefully be given in IMF programs to accompanying budgetary stabilization with efficiency-improving measures in productive sectors (especially public enterprises). This is essential if their reduced dependency on budgetary transfers is to become long-term sustainable reality.

Similar considerations apply to the design of policies in the foreign trade sector. The IMF tends to emphasize exchange rate action and management and to accord less attention to the details of trade policy reform such as export promotion, import liberalization, and tariff policy. Here again there is scope for more detailed program design, notably to ensure that trade policy complements budgetary policy. The Morocco experience is particularly illuminating in this regard. Failure to press hard enough for reform of indirect taxes can compromise trade policy reform and so the growth of a country's capacity to earn foreign exchange.

The World Bank's Approach to Stabilization and Adjustment

Prior to 1979-80, the Bank's lending operations consisted exclusively of specific projects and investment lines of credit to financial intermediaries. The first attempt at adjustment lending was the 1980 structural adjustment loan (SAL), which was based on the 1979 *Basic Economic Report*.

This report had correctly identified the need for macroeconomic stabilization (higher savings and lower consumption, both public and private), as well as higher exports. Otherwise, sharply rising external debt service was expected to compromise prospects for growth and poverty reduction. In addition, the need to reform price policy and improve production incentives, especially for exports and agriculture, was strongly emphasized. In the light of this report and the rather poor economic performance from 1973 to 1977, the Moroccan government and the World Bank initiated, in late 1979, a joint research study of production and investment incentives in the industrial sector. Because its results were not yet available, the 1980 SAL was based only on the *Basic Economic Report*. However, it proved impossible to translate the report's analysis into a coherent adjustment program with well-defined conditionalities related to specific, workable structural reforms. In fact, the principal conditionalities that were proposed covered project selection criteria for public sector investments; public enterprise reform, including the elimination of operating and capital transfers from the Treasury; a requirement that public enterprise investments be financed at least 30 percent from retained earnings; and the introduction of positive real interest rates. But the SAL was too diffused, and it was prepared too quickly for a consensus to develop either in the Bank or Morocco around the proposed conditionalities, important as they may have been for public enterprise reform. Many of its recommendations were for studies rather than actions. The proposals for project selection encountered strong resistance from the government, and eventually the Bank's management decided that the proposed conditionalities were too weak. Consequently, negotiations were suspended and the SAL abandoned.

Rather than attempt to revive the SAL, the Bank decided to approach Morocco's needs for far-reaching structural reforms through a series of sectoral adjustment operations in industry, finance, education, agriculture, and public enterprises. It was felt that the sectoral approach would (a) be politically more acceptable to the government than the SAL's across-the-board approach; and (b) allow deeper preparatory analysis in each sector. An important part of the strategy was to proceed only where the preparatory analysis was completed. The first two operations focused on Industry and Trade Policy Adjustment (ITPA) and were based primarily on the output of the joint research project and follow-up Bank missions, especially in the financial sector. The project's conclusions emphasized the need to reform the structure of production incentives for exports and the domestic market in order to create a policy environment that would justify an expansion of traditional projects and investment lines of credit. The premise of the ITPA operations was that it was essential to increase Morocco's supply of foreign exchange through economically efficient exports and import substitution and to enhance employment generation. The corresponding adjustment program then had to address the following issues:

- reduction of antiproduction bias, especially for exports but also for the domestic market, that had resulted from the appreciation of the exchange rate and the pervasive price controls of the preceding decade;

- reduction of anti-export bias resulting from export taxes and licenses, tariff and nontariff protection of the domestic market, cumbersome international trade procedures, and ill-adapted financial policies;

- reduction of excessive dispersion of the structure of effective protection on the domestic market, and associated rent seeking;

- reduction of bias against financial savings and their mobilization by financial intermediaries;

- reduction of anti-employment bias in production and investment incentives.

These issues had all been clearly identified by the research project and follow-up Bank missions. The collaborative nature of the work had resulted in a consensus within the business community and the Moroccan government (particularly the Ministry of Industry, Ministry of Economic Affairs, and to a lesser extent the Ministry of Finance and the Bank of Morocco) that these were indeed the critical issues. During implementation of the adjustment program other issues were addressed that related primarily to public enterprises as well as to the arrears problems and expenditure control problems identified by the International Monetary Fund during the preparation of its standby operations.

To redress this situation, the ITPA program proposed a series of measures to change the structure of incentives at the microeconomic level, but it took due account of their macroeconomic consequences—particularly regarding the budget. As explained at length in Chapter 5, the key components of the program were

- a compensated devaluation in which the exchange rate change is accompanied by a reduction in import levies, coupled with other export promotion and trade simplification measures to reduce anti-export bias (including elimination of export monopolies and licensing);

- a phased and sustainable import liberalization program to spur efficiency in firms selling on the domestic market;

- a phased tariff reform (undertaken in conjunction with the import liberalization program and flexible exchange rate management) to further reduce tariff-induced anti-export bias and dispersion of effective protection on the domestic markets;

- compensating increases in indirect taxes and transfers from the phosphate company to make up revenue losses caused by lower import levies;

- deregulation of domestic prices by allowing producers to revise prices without prior authorization, except for certain categories of basic, generally subsidized, commodities;

- higher interest rates to encourage worker remittances and financial savings and to discourage unnecessarily capital-intensive technologies; as well as a reform of bank reserve and placement requirements to spur mobilization of financial savings by financial intermediaries.

Particular importance was given to the issue of recovering the revenue losses from lower import levies by increasing the manufacturer's sales taxes and by raising transfers from the state phosphate company. Both measures, particularly regarding internal taxes, were considered politically feasible on the basis of the 1982 and 1983 moves to raise revenues through internal taxes rather than import duties in line with the Ministry of Industry's desire to reduce anti-export bias. It was also expected that the manufacturer's sales tax would be transformed into a value added tax and extended to the distribution sector in the course of the IMF standbys. Explicit consideration was given by the Ministry of Industry to reducing direct tax holidays to exporters and new firms, but these ideas were not enacted into law until 1988. Similarly, in revising the investment code in 1983, the Ministry of Industry successfully advocated the use of employment subsidies per job created, in order to encourage firms to expand output at the margin by increasing employment rather than investing in new plant and equipment. Finally, a key characteristic of the ITPA program was the decision by the Ministry of Industry to discuss the reforms openly with the industrial sector, which was treated as a partner in the preparation process. As a result, the firms had time to prepare for the adjustments.

In terms of the methodology developed in Chapter 3, the ITPA approach to adjustment and to improving the current account deficit stressed the need to increase exports of goods and services as well as to augment worker remittances. In this connection, Chapter 5 has analyzed the beneficial impact on the external current account of improved foreign exchange earnings from manufactured exports, tourism, and worker remittances (see Figures 5.1 to 5.3). An additional consequence of this growth has been reduced volatility of export earnings; phosphate and agricultural products no longer dominate exports as they once did (see Figure 5.3). Reducing imports was emphasized only insofar as there was scope for efficient import substitution.

The ITPA program can, of course, be interpreted in terms of the model that represents the current account surplus or deficit as the sum of the overall budget surplus or deficit ($S_g - I_g$) and financial saving or dissaving in the nongovernment sector ($S_{ng} - I_{ng}$). The argument would be that the primary purpose of the ITPA program was to raise GNP, and so nongovernment savings (S_{ng}), through higher exports and worker remittances as well as lower imports competing from economically efficient import substitution. However, export growth and some categories of import substitution in industry and agriculture were being significantly disadvantaged by the overvalued exchange rate and structure of tariff and nontariff protection. The key to redressing anti-export and antiproduction bias was replacing tariff protection of the domestic market with exchange rate promotion via a devaluation and reduction of import levies. However, this strategy entailed a loss of import duty revenues and hence an adverse impact on the budget deficit ($S_g - I_g$). Therefore, offsetting revenue-raising measures were needed in the form of higher internal indirect taxes, as well as export taxes on goods where Morocco had market power—namely, phosphates. The need for these complementary fiscal revenues had been stressed since 1979-80 as part of the joint World Bank-Government of Morocco research project.[3]

In other words, the ITPA program's macroeconomic objectives were to improve the external current account by increasing output, and thus nonbudgetary savings (S_{ng}), while minimizing any short-run offsetting increase in the budget deficit ($S_g - I_g$). Additional improvement in the current account was anticipated from

---

3. The desirability of reducing trade taxes has recently been stressed by the Fiscal Affairs Department of the IMF (1988). Written more than ten years after the beginning of the incentives research project, the IMF paper also stresses that trade tax reduction is unlikely to be feasible unless accompanied by a reform of internal indirect taxes.

increasing interest rates. Higher rates were expected to raise worker remittances and encourage firms to substitute labor for capital, so reducing the rate of capital formation (I_{ng}) at the margin without, however, compromising overall growth. Figures 2.1a and 2.2 suggest some measure of success. Net financial savings in the nongovernment sector increased substantially during the ITPA program. They made as significant an increase in improving the external current deficit as did the improvements in the budget deficit (see Figure 2.1a).

Compared with the IMF standbys, the ITPA program had a more substantive microeconomic component. Whereas the IMF standbys emphasized the need to depreciate the exchange rate, ITPA stressed that depreciation should be accompanied by a change in the structure of relative production incentives in favor of exports. Similarly, the ITPA program emphasized the need to improve the efficiency of resource allocation at the firm level by exposing firms to the rivalry of foreign competition even on the domestic market. To encourage exports, expose firms to import competition on the domestic market, and change the relative structure of incentives in favor of exports, the government needed to pursue a combination of trade policy measures rather than just the simple exchange rate depreciation advocated by the IMF. In addition to depreciation, a phased import liberalization program coupled with tariff reduction was required.

Over the long run, the progressive replacement of tariff protection by exchange rate promotion was expected to enhance the competitiveness of domestically produced inputs that could be substituted for imported raw materials and intermediates used by exporters. In this way, the ITPA program aimed at augmenting the domestic content of locally manufactured exports. Similarly, interest rate reforms would, it was hoped, spur worker remittances, mobilize additional financial savings, and reduce bias against labor-intensive production technologies.

The ITPA program did not view export promotion as a simple matter of getting the prices right. It paid equal attention to the overall economic and administrative environment for exports. It recommended improvements in the areas of export financing and insurance, customs and tax regimes for direct and indirect exporters, and the simplification of international trade procedures. Neither was reform of the domestic regulatory framework neglected, especially regarding price control.

Another important component of the ITPA process was that the basic research had been conducted within the Ministry of Industry over a three-year period. Moreover, it was realized that demand compression was unavoidable, which made the authorities willing to look seriously at supply-side structural reforms that would mitigate the adverse impact of stabilization. Thus, when crisis struck in 1983 the Moroccan authorities had a domestically generated diagnosis of the problems besetting the industrial sector. The collaborative nature of the research project, which was based on a firm-level survey, made the definition of a politically sustainable adjustment program much easier. An equally important factor was the open dialogue between the government and industrial firms. The Moroccan authorities made considerable efforts to persuade industrialists that the purpose of the reforms was to strengthen, rather than weaken, the industrial sector. The government indicated its willingness to reverse itself if concrete evidence indicated that firms were suffering material injury from duly documented unfair competition. Another key step was to give firms at least one year to prepare for import liberalization.

The ITPA program paid considerable attention to budgetary questions. Since reducing anti-export bias implied devaluation and import duty reduction, the program had to have a major budgetary component. The shortfall in import duty receipts, including from the special import tax, required offsetting increases in internal taxes (both direct and indirect), higher budgetary transfers from the phosphate company, as well as decreases in expenditures. In that regard, the ITPA program clearly identified the potential problem and proposed a practical solution that was based on the successful process, begun in 1982-83, of raising additional revenues through higher indirect taxes rather than higher trade taxes. Yet, during the ITPA program's implementation, insufficient pressures were brought on the Ministry of Finance to continue this process by raising the rates of the VAT when it was introduced in 1986. These upward adjustments were all the more needed since the VAT reform entailed an additional revenue loss because of the generalization of deductions. The public finance situation was further exacerbated by difficulties on the expenditure side. Expenditures, especially once committed, had proved exceedingly difficult to cut, even if resources were not available for payment. In fact, reduced cash expenditures often implied an equivalent buildup of arrears, thus increasing the need for budgetary resources in the future. And, finally, depressed prices in the world phosphate market coupled with a large investment program severely constrained budgetary transfers from the phosphate company. As a result of these factors, the pace of reduction of trade taxes had to be slowed down in 1985, was effectively arrested in 1986, and was effectively reversed in January 1988.

This sequence of events indicates to what extent trade policy reform can fall prey to budgetary developments. Trade policy adjustment programs must take better account of these budgetary constraints than was possible in Morocco—despite the considerable efforts made in that regard by the ITPA loans and the subsequent Public Enterprise Restructuring Loan (PERL) that specifically addressed the arrears problem.

Conclusion

The IMF and the World Bank approaches to Morocco's economic difficulties had the same macroeconomic objective—namely, to bring the external current account into line with the available supply of foreign savings. Yet the means to this end were quite different. The Bank concentrated primarily on augmenting the supply of foreign exchange and improving the efficiency of resource allocation. The IMF emphasized expenditure containment in general and reduction of the budget deficit in particular. The Fund was also instrumental in persuading the country's external creditors to reschedule principal and interest on outstanding debt. This significantly eased Morocco's external financing problems, as Figure 2.8 shows.

The challenge has been to maximize the complementarity of the two institutions' approaches. This is the nexus of the stabilization adjustment debate, since enhancing the conditions for economic growth requires macroeconomic stability, a larger supply of foreign exchange, and greater efficiency of resource allocation. The Moroccan case shows that achieving complementarity is difficult, both technically and politically. It requires the creation of consensus externally (between the Bank and the Fund) and internally (between the Ministry of Finance, the Central Bank, operational ministries, and the industrial sector).

Achieving complementarity may require particular attention in three areas, inter alia: indirect tax policy, public expenditure policy, and public sector arrears. Budgetary stabilization attained by increasing import duties and levies may increase anti-export bias. At the same time, higher growth requires an increase in the supply of foreign exchange and hence a reduction in anti-export bias. This may be achieved in part by lowering incentive-distorting import levies. The potential conflict between reducing the budget deficit and attenuating anti-export bias may be avoided if the reform of production incentives is accompanied by a reform of indirect taxes. Import levies should be replaced by internal taxes, both direct and indirect, that do not affect the relative levels of production incentives for exports and the domestic market. The Moroccan case shows that it is not enough for the Bank to prepare trade policy reforms on the assumption that such a tax reform will materialize. It must be an active proponent of, and participant in, the tax reform process itself.

The reduction of public expenditure may also be an essential element of a consistent stabilization cum adjustment program. Expenditure reduction must be planned so as to minimize adverse effects on the competitiveness of productive sectors. For example, reducing budgetary transfers to productive enterprises, while desirable as a means of reducing the budget deficit, should be accompanied by measures to increase the efficiency of the enterprises in question. In this way, an offsetting decline in productive sector savings (that is, nonbudgetary savings) can be avoided. Otherwise, reducing budgetary transfers will have no net effect on the external current account. Likewise, reduction in government capital expenditures should minimize, as far as possible, any adverse impact on the supply of public infrastructure or government services that are used in the production of exports and import substitutes.

It is also essential that government and public enterprise arrears be capped and then reduced, since their continued accumulation can choke off the supply of working capital credit to firms holding the unpaid claims on government and public enterprises. This eventuality can easily have a pronounced downward impact on capacity utilization. Consequently, effectively attenuating the arrears problem should have a very high priority in structural adjustment programs.

Adjustment programs designed to increase the supply of foreign exchange must do much more than simply improve relative prices and production incentives for exports and the domestic market. Considerable attention also must be given to improving the overall regulatory and administrative environment relating to all aspects of international trade and related procedures. The same is true of the domestic regulatory framework relating to price controls, the structure of competition, barriers to entry and exit, and labor market regulations. In the Moroccan case, most attention was given to liberalizing price controls, except for subsidized commodities and those produced by completely protected monopolists.

Finally, this case study—as well as a recent favorable evaluation of the ITPA program (World Bank, 1989)—indicate the importance of getting the politics right in at least two regards. First, the joint, in-depth analysis of the incentives structure by the World Bank and the Moroccan government, on the basis of a firm-level survey, led to a consensus within Morocco that adjustment was needed. As a result, the adjustment program was largely designed and managed in Morocco, although the Bank actively participated in the initial stages. The Bank played a particularly important role in trying to reconcile the divergent positions of different participants in the design process. Second, the Moroccan experience points to the benefits from openly discussing the reform program with the principal actors likely to be affected by the process. This openness may have made the overall pace of reform slower than elsewhere. But, with the exception of the entirely avoidable January 1988 tariff increase, it meant that the adjustment program had been viewed by the industrial sector as sustainable and credible. In large part, this was because the government went to considerable lengths to persuade the industrial sector that the purpose of the reforms was to make firms better able to compete, both in terms of price and quality, against fairly priced international competition. At the same time, the government stressed its willingness to modify its policy on trade liberalization should the survival of the industrial sector be threatened by unfair trading practices on the part of its commercial partners. As a result, Moroccan industrialists have felt that the government has been a pragmatic rather than dogmatic advocate of essential economic reforms.

References

Balassa, Bela. 1988a. *Public Finance and Economic Development*. Policy, Planning and Research, Staff Working Paper 31. Washington, D.C.: The World Bank.

_____. 1988b. *A Conceptual Framework for Adjustment Policies*. Policy, Planning and Research, Staff Working Paper 139. Washington, D.C.: The World Bank.

Farhadian, Ziba, and Lorie and Menachem Katz. 1988. *Fiscal Dimensions of Trade Policy*. Fiscal Affairs Department. Washington, D.C.: The World Bank. May.

Horton, Brendan, and Garry Pursell. 1984. *Incentives and the Efficiency of Resource Allocation in the Industrial Sector*. Unpublished. Washington, D.C.: The World Bank.

International Monetary Fund. various years. *International Financial Statistics*.

Krueger, Anne. 1978. *Liberalization Attempts and Consequences*. Cambridge, Mass.: Ballinger.

Linn, Johannes, and Deborah Wetzel. 1989. *Public Finance, Trade and Development: What Have We Learned?* Policy, Planning and Research, Staff Working Paper 181. Washington, D.C.: The World Bank.

Morocco, Government of. 1983a. *Rapport sur la Structure des Incitations, Rapport de Synthèse et Rapports Sectoriels*. Rabat. Mars.

_____. 1983b. *Note Relative aux Aspects Fiscaux de l'Adjustment de l'Economie*. Ministry of Commerce and Industry. Rabat. Septembre.

_____. 1983c. *Note Relative à la Structure des Recettes Fiscales*. Ministry of Commerce and Industry. Rabat. Juin.

_____. 1986. *La Politique des Prix et d'Incitations dans le Secteur Agricole*. Rapport final. Rabat. Janvier.

World Bank. 1979. *Morocco: Basic Economic Report*. Washington, D.C.

_____. 1981. *Morocco: Structural Adjustment Loan*. Staff Appraisal Report. Unpublished. Washington, D.C.

_____. 1983. *The Industrial Incentive System in Morocco*. RPO 671-85. Washington, D.C.

_____. 1984a. *Morocco: Industrial Incentives and Export Promotion*. Washington, D.C. January.

_____. 1984b. *Morocco: Industrial Trade Policy Adjustment Loan, Report and Recommendations of the President*. Washington, D.C. January.

_____. 1985a. *Morocco: Second Industrial Trade Policy Adjustment Loan, Report to the President*. Washington, D.C. July.

_____. 1985b. *Morocco: Financial Sector Report*. Washington, D.C. November.

_____. 1986. *Morocco: Agricultural Prices and Incentives Study*. Washington, D.C. January.

_____. 1987a. *Morocco: Second Agriculture Sector Adjustment Loan, Report and Recommendations of the President*. Washington, D.C.

_____. 1987b. *Morocco: Public Finance and Economic Growth*. Washington, D.C. September.

_____. 1988a. *Morocco: The Impact of Liberalization on Trade and Industrial Adjustment*. Washington, D.C. March.

_____. 1988b. *Morocco: Structural Adjustment Loan, Report and Recommendations of the President*. Washington, D.C.

_____. 1989. *Program Performance and Audit Report: Industrial Trade Policy Adjustment Loans*. Operations Evaluation Department. Washington, D.C. June.

Annexes

Annex 1
Synopsis of Balance of Payments Support, 1970-87

International Monetary Fund

The IMF finances temporary balance of payments disequilibria using its own (that is, ordinary) and borrowed resources, which are of a revolving character. The Fund transfers these resources to its members in the form of special drawing rights (SDRs) under a variety of facilities (see Table 1). Such transfers are generally referred to as *purchases*. The member must repay the IMF within three to seven years, although certain facilities extend the repayment period to ten years from the date of borrowing. Repayments are usually called *repurchases*.

Table A.1 Main IMF Facilities

Facility	*Eligibility criteria*	*Years to repay*
Compensatory Financing Facility (CFF)	Shortfall of export earnings or excess cereals imports	3-5
Credit Tranche Facilities (standbys)	General balance of payments difficulties	3-5
Extended Fund Facility (EFF)	Balance of payments difficulties resulting from structural problems	3-10

In general, drawings under the standby and extended facilities are made available after successful periodic reviews. These reviews indicate that quantitative performance criteria have been satisfied and, if necessary, understandings reached about the future course of certain policy variables. Drawings under the compensatory financing facility usually are not conditional in this manner. The repayment terms of particular operations may vary, depending on whether the IMF resources are ordinary or borrowed.

The IMF tries to avoid detailed discussions on the evolution and management of a country's policies and instruments. This is the sovereign responsibility of national policymakers, within the constraints of the legally binding program performance criteria. Programs may include indicative nonbinding targets for receipts, expenditures, and the ratio of current account and Treasury deficits to GDP.

Table 2 summarizes purchases and repayments until the end of 1986. Operations before 1975 have been aggregated. Thus, by the end of 1986, the International Monetary Fund had lent nearly US$1.9 billion to Morocco; about 5.5 percent of this sum had been loaned before 1975.[1] Since 1976, total use of Fund resources has amounted to just over US$1.75 billion.

Table A.2 IMF Operations in Morocco, 1970-86 (millions of SDRs and US dollars)

Year	Purchases (millions of SDR)	Repurchases	$/SDR	Purchases (millions of U.S. dollars)	Repurchases	Net drawings
1945-57	91.4	91.4	1.2142	111.0	111.0	
1976	143.7		1.1545	165.9		165.9
1978	56.0		1.2520	70.1		70.1
1979		23.2	1.2920		30.0	-30.0
1980	184.5	67.4	1.3015	240.1	87.7	152.4
1981	192.8	53.4	1.1792	227.3	62.9	164.4
1982	433.7	32.5	1.1040	478.8	35.9	442.9
1983	134.8	23.3	1.0690	144.1	24.9	119.2
1984	180	47.5	1.0250	184.5	48.7	135.8
1985	215.1	143.5	1.0153	218.4	145.4	72.7
1986	30.0	274.5	1.1732	35.2	321.7	-286.5
Total	1,661.5	755.9		1,869.9	862.5	1,007.4

Source: IMF and author's estimates.

These resources were lent to Morocco in the context of the operations summarized in Table 3. The first two columns specify the facility used and the date, and the next two show the amount of resources made available in SDRs and dollars. The last column indicates whether the resources were partly or fully used and in the latter case the month and year when the underlying agreement with the Fund was cancelled.

From 1976 to 1980, Morocco made purchases under the Gold and First Credit Tranche Facilities, the Oil Facility, and the Compensatory Financing Facility (CFF). Since then, the drawings have been under the Extended Fund Facility, standby programs, and the CFF. The biggest program was the 1980 EFF, but this was cancelled in December 1981 after only 29.6 percent of the allowable purchases had been made. The same fate befell the 1985 standby.

---

1. The low 1986 drawing (SDR 30 million) corresponded to the first made under the December 1986 standby.

Table A.3 IMF Operations in Morocco by Type, 1970-86 (millions of SDRs and US dollars)

Nature of operation	Date	Millions of SDRs	US$	Status
Gold Tranche	1/76	28	32	100% used
First Credit Tranche	4/76	41	47	100% used
Compensatory Financing Facility (CFF)	4/76	57	66	100% used
Oil Facility	5/76	18	21	100% used
Extended Fund Facility	8/80-7/83	810	1,054	Cancelled 12/81; 240 M.SDR used
Standby	4/82-7/83	282	311	Completed
CFF	4/82	236	261	100% used
Standby	9/83-2/85	300	321	Completed
Standby	9/85-2/87	225	228	Cancelled 12/85; 10 M.SDR used
CFF	9/85	115	117	100% used
Standby	12/86-2/88	230	270	In force

World Bank

In keeping with its vocation as a development bank, the World Bank before 1980 made only project specific investment loans in Morocco. Since then, the Bank has shifted its emphasis to lending in support of policy reform. Initially, this took the form of the unsuccessful structural adjustment loan. Subsequently, the SAL approach was abandoned in favor of a series of sector adjustment loans beginning in 1984 (see Table 4). The total amount of Bank resources committed to the support of policy reform has therefore reached about $1 billion. Repayment terms are generally seventeen years with four years grace.

Table A.4 World Bank Operations in Morocco

Name of operation	Date of effectiveness	Amount in millions of US$	Status
Structural Adjustment Loan (SAL)		500.0	Negotiations suspended in 1981
Industry & Trade Policy Adjustment 1 (ITPA1)	January 1984	150.4	Both tranches released on schedule
Industry & Trade Policy Adjustment 2 (ITPA2)	July 1985	200.0	Second tranche released after a six-month delay
Agriculture Structural Adjustment 1 (ASAL1)	October 1985	100.0	Fully disbursed on schedule
Agriculture Structural Adjustment 2 (ASAL2)	June 1988	200.0	Negotiated August 1987
Education 1	September 1986	150.0	Fully disbursed on schedule
Public Enterprise Restructuring Loan (PERL)	December 1987	240.0	In force

Annex 2
Tables

Table 1 External Current Account After Debt Relief, Budget Deficit, and Net Nongovernmental Financial Savings, Including and Excluding Arrears, 1970-87

Year	Gross domestic product	External current account	Budget deficit payment order basis	Net nongovt. financial savings payment order basis	Budget deficit, cash basis inc. change in arrears	Net nongovt. savings, cash basis inc. change in arrears	Money Supply
	GDP	CA	Sg-Ig	Sng-Ing	S'g-I'g	S'ng-I'ng	M2

A. In billion Dh

1970	20.02	-0.63	-0.61	-0.02	-0.61	-0.02	5.95
1971	22.00	-0.30	-0.66	0.36	-0.66	0.36	6.69
1972	23.34	-0.22	-0.94	0.72	-0.93	0.71	7.89
1973	25.64	0.43	-0.52	0.95	-0.51	0.94	9.19
1974	33.60	1.04	-1.32	2.35	-1.32	2.36	11.85
1975	36.38	-2.21	-3.46	1.25	-3.46	1.25	14.28
1976	41.01	-5.99	-7.42	1.43	-7.42	1.43	16.90
1977	49.76	-8.22	-7.88	-0.34	-7.89	-0.33	20.05
1978	55.15	-5.62	-6.10	0.48	-6.23	0.61	24.31
1979	62.04	-5.97	-6.31	0.34	-6.31	0.34	27.79
1980	70.16	-5.59	-7.44	1.85	-6.67	1.08	30.83
1981	76.74	-9.63	-11.35	1.72	-10.78	1.15	35.60
1982	90.09	-11.43	-11.52	0.09	-8.56	-2.87	38.96
1983	94.63	-6.21	-11.99	5.78	-11.41	5.20	45.74
1984	104.83	-8.79	-12.56	3.77	-9.06	0.27	50.47
1985	119.66	-8.64	-12.46	3.82	-11.05	2.41	59.42
1986	134.33	-1.96	-8.23	6.27	-8.78	6.82	68.85
1987	140.00	1.37	-9.05	10.42	-9.67	11.04	

B. As percentage of GDP

1970	20.02	-3.2	-3.0	-0.1	-3.0	-0.1	29.7
1971	22.00	-1.4	-3.0	1.6	-3.0	1.6	30.4
1972	23.34	-0.9	-4.0	3.1	-4.0	3.1	33.8
1973	25.64	1.7	-2.0	3.7	-3.0	3.7	35.9
1974	33.60	3.1	-3.9	7.0	-3.9	7.0	35.3
1975	36.38	-6.1	-9.5	3.4	-9.5	3.4	39.2
1976	41.01	-14.6	-18.1	3.5	-18.1	3.5	41.2
1977	49.76	-16.5	-15.8	-0.7	-15.9	-0.7	40.3
1978	55.15	-10.2	-11.1	0.9	-11.3	1.1	44.1
1979	62.04	-9.6	-10.2	0.6	-10.2	0.6	44.8
1980	70.16	-8.0	-10.6	2.6	-9.5	1.5	43.9
1981	76.74	-12.5	-14.8	2.2	-14.0	1.5	46.4
1982	90.09	-12.7	-12.8	0.1	-9.5	-3.2	43.2
1983	94.63	-6.6	-12.7	6.1	-12.1	5.5	48.3
1984	104.83	-8.4	-12.0	3.6	-8.6	0.3	48.1
1985	119.66	-7.2	-10.4	3.2	-9.2	2.0	49.7
1986	134.33	-1.5	-6.1	4.7	-6.5	5.1	51.3
1987	140.00	1.0	-6.5	7.4	-6.9	7.9	0.0

Source: World Bank and author's estimates.

Table 2 External Current Account, Resource Balance, and Net Factor Services and Transfers from Abroad

Year	Gross domestic product	Export of goods and nonfactor services X	Imports of goods and nonfactor services M	Resource balance = exports less imports X-M	Factor services and transfers from abroad, net FSTn	External current account CA
A. In billion Dh						
1970	20.02	3.53	4.32	-0.79	0.16	-0.63
1971	22.00	3.73	4.34	-0.61	0.31	-0.30
1972	23.34	4.34	4.49	-0.15	0.37	0.21
1973	25.64	5.34	5.67	-0.33	0.76	0.43
1974	33.60	9.24	9.45	-0.21	1.25	1.04
1975	36.39	8.18	12.14	-3.96	1.75	-2.21
1976	41.01	7.59	15.56	-7.97	1.98	-5.99
1977	49.76	8.41	18.57	-10.16	1.94	-8.22
1978	55.15	9.03	16.58	-7.56	1.94	-5.62
1979	62.04	10.55	18.53	-7.97	2.01	-5.97
1980	70.16	12.88	20.66	-7.77	2.18	-5.59
1981	76.74	15.94	27.47	-11.53	1.90	-9.63
1982	90.09	17.88	31.31	-13.43	2.00	-11.44
1983	94.63	21.29	30.03	-8.74	2.53	-6.21
1984	104.81	26.77	38.68	-11.91	3.12	-8.79
1985	119.66	32.05	43.69	-11.64	3.00	-8.64
1986	134.33	32.92	42.72	-9.80	7.83	-1.97
1987	140.00	35.40	42.32	-6.92	8.29	1.37
B. As percentage of GDP						
1970	20.02	17.6	21.6	-3.9	0.8	-3.2
1971	22.00	17.0	19.7	-2.8	1.4	-1.4
1972	23.34	18.6	19.2	-0.7	1.6	0.9
1973	25.64	20.8	22.1	-1.3	3.0	1.7
1974	33.60	27.5	28.1	-0.6	3.7	3.1
1975	36.39	22.5	33.4	-10.9	4.8	-6.1
1976	41.01	18.5	37.9	-19.4	4.8	-14.6
1977	49.76	16.9	37.3	-20.4	3.9	-16.5
1978	55.15	16.4	30.1	-13.7	3.5	-10.2
1979	62.04	17.0	29.9	-12.9	3.2	-9.6
1980	70.16	18.4	29.4	-11.1	3.1	-8.0
1981	76.74	20.8	35.8	-15.0	2.5	-12.5
1982	90.09	19.8	34.8	-14.9	2.2	-12.7
1983	94.63	22.5	31.7	-9.2	2.7	-6.6
1984	104.81	25.5	36.9	-11.4	3.0	-8.4
1985	119.66	26.8	36.5	-9.7	2.5	-7.2
1986	134.33	24.5	31.8	-7.3	5.8	-1.5
1987	140.00	25.3	30.2	-4.9	.59	1.0

Source: World Bank and author's estimates.

Table 3 External Current Account, Domestic Savings, Investment, and Net Factor Services and Transfers from Abroad

Year	Gross domestic product	Domestic savings Sd	Investment I	Domestic savings less investment Sd-I	Factor services and other transfers, net FSTn	External current account CA
A. In billion Dh						
1970	20.02	2.91	3.70	-0.79	0.16	-0.63
1971	22.00	3.34	3.95	-0.61	0.31	-0.30
1972	23.34	3.42	3.57	-0.15	0.37	0.21
1973	25.64	4.00	4.32	-0.33	0.76	0.43
1974	33.60	6.71	6.92	-0.21	1.25	1.04
1975	36.39	5.21	9.17	-3.96	1.75	-2.21
1976	41.01	3.56	11.53	-7.97	1.98	-5.99
1977	49.76	6.87	17.03	-10.16	1.94	-8.22
1978	55.15	6.47	14.03	-7.56	1.94	-5.62
1979	62.04	7.22	15.20	-7.97	2.01	-5.97
1980	70.16	8.10	15.87	-7.77	2.18	-5.59
1981	76.74	5.66	17.19	-11.53	1.90	-9.63
1982	90.09	7.54	20.97	-13.43	2.00	-11.44
1983	94.63	11.02	19.76	-8.74	2.53	-6.21
1984	104.81	10.88	22.82	-11.94	3.12	-8.81
1985	119.66	14.36	27.38	-13.02	3.090	-10.03
1986	134.33	17.46	27.26	-9.80	7.83	-1.97
1987	140.00	19.78	26.70	-6.92	8.29	1.37
B. As percentage of GDP						
1970	20.02	14.5	18.5	-3.95	0.8	-3.2
1971	22.00	15.2	17.9	-2.77	1.4	-1.4
1972	23.34	14.6	15.3	-0.65	1.6	0.9
1973	25.64	15.6	16.9	-1.27	3.0	1.7
1974	33.60	20.0	20.6	-0.63	3.7	3.1
1975	36.39	14.3	25.2	-10.88	4.8	-6.1
1976	41.01	8.7	28.1	-19.43	4.8	-14.6
1977	49.76	13.8	34.2	-20.43	3.9	-16.5
1978	55.15	11.7	25.4	-13.70	3.5	-10.2
1979	62.04	11.6	24.5	-12.85	3.2	-9.6
1980	70.16	11.5	22.6	-11.08	3.1	-8.0
1981	76.74	7.4	22.4	-15.02	2.5	-12.5
1982	90.09	8.4	23.3	-14.91	2.2	-12.7
1983	94.63	11.6	20.9	-9.24	2.7	-6.6
1984	104.81	10.4	21.8	-11.39	3.0	-8.4
1985	119.66	12.0	22.9	-10.88	2.5	-8.4
1986	134.33	13.0	20.3	-7.30	5.8	-1.5
1987	140.00	14.1	19.1	-4.94	5.9	1.0

Source: World Bank and author's estimates.

Table 4 External Current Account, National Savings and Investments

Year	Gross domestic product	Domestic savings Sd	Factor services plus transfers, net FSTn	National savings Sn = Sd+FSTn	Investment I	External current account Sn-I
A. In billion Dh						
1970	20.02	2.91	0.16	3.07	3.70	-0.63
1971	22.00	3.34	0.31	3.65	3.95	-0.30
1972	23.34	3.42	0.37	3.78	3.57	0.21
1973	25.64	4.00	0.76	4.76	4.32	0.43
1974	33.60	6.71	1.25	7.96	6.92	1.04
1975	36.39	5.21	1.75	6.95	9.17	-2.21
1976	41.01	3.56	1.99	5.54	11.53	-5.99
1977	49.76	6.87	1.94	8.81	17.03	-8.22
1978	55.15	6.47	1.94	8.41	14.03	-5.62
1979	62.04	7.22	2.01	9.23	15.20	-5.97
1980	70.16	8.10	2.18	10.28	15.87	-5.59
1981	76.74	5.66	1.90	7.56	17.19	-9.63
1982	90.09	7.54	2.00	9.53	20.97	-11.44
1983	94.63	11.02	2.53	13.55	19.76	-6.21
1984	104.81	10.88	3.12	14.01	22.82	-8.81
1985	119.66	14.36	3.00	17.35	27.38	-10.03
1986	134.33	17.46	7.83	25.29	27.26	-1.97
1987	140.00	19.78	8.29	28.07	26.70	1.37
B. As percentage of GDP						
1970	20.02	14.5	0.8	15.3	18.5	-3.2
1971	22.00	15.2	1.4	16.6	17.9	-1.4
1972	23.34	14.6	1.6	16.2	15.3	0.9
1973	25.64	15.6	3.0	18.5	16.9	1.7
1974	33.60	20.0	3.7	23.7	20.6	3.1
1975	36.39	14.3	4.8	19.1	25.2	-6.1
1976	41.01	8.7	4.8	13.5	28.1	-14.6
1977	49.76	13.8	3.9	17.7	34.2	-16.5
1978	55.15	11.7	3.5	15.2	25.4	-10.2
1979	62.04	11.6	3.2	14.9	24.5	-9.6
1980	70.16	11.5	3.1	14.6	22.6	-8.0
1981	76.74	7.4	2.5	9.8	22.4	-12.5
1982	90.09	8.4	2.2	10.6	23.3	-12.7
1983	94.63	11.6	2.7	14.3	20.9	-6.6
1984	104.81	10.4	3.0	13.4	21.8	-8.4
1985	119.66	12.0	2.5	14.5	22.9	-8.4
1986	134.33	13.0	5.8	18.8	20.3	-1.5
1987	140.00	14.1	5.9	20.1	19.1	1.0

Source: World Bank and author's estimates.

Table 5 Total Investment, Government Investment (payment order and cash basis), Nongovernment Fixed Investment, and Variation in Stocks

Year	Gross domestic product	Govt. and nongovt. investment Total	Government payment order basis Ig	Non-government Ing	Government investment Cash basis	Financed by arrears	Nongovt. investment Change in stocks	Fixed investment Ingf
A. In billion Dh								
1970	20.02	3.70	1.15	2.56	1.15	0.00	0.71	1.84
1971	22.00	3.95	1.12	2.83	1.12	0.00	0.68	2.15
1972	23.34	3.57	1.19	2.38	1.19	0.00	0.39	1.99
1973	25.64	4.32	1.21	3.12	1.21	0.00	0.85	2.26
1974	33.60	6.92	2.24	4.69	2.24	0.00	1.99	2.70
1975	36.39	9.17	4.45	4.71	4.45	0.00	0.13	4.58
1976	41.01	11.53	8.12	3.41	8.12	0.00	-0.64	4.06
1977	49.76	17.03	10.31	6.72	10.31	0.00	1.13	5.60
1978	55.15	14.03	6.63	7.40	6.63	0.00	0.30	7.10
1979	62.04	15.20	9.01	6.19	9.01	0.00	0.32	5.87
1980	70.16	15.87	8.07	7.80	5.83	2.24	1.06	6.74
1981	76.74	17.19	9.32	7.87	7.00	2.32	0.36	7.51
1982	90.09	20.97	12.09	8.88	7.61	4.48	-0.12	9.00
1983	94.63	19.76	9.96	9.80	5.71	4.25	-0.79	10.58
1984	104.81	22.82	8.86	13.96	3.76	5.10	0.36	13.60
1985	119.66	27.38	8.86	18.52	3.06	5.80	2.41	16.10
1986	134.33	27.26	5.52	21.74	2.40	3.12	1.00	20.74
1987	140.00	26.70	9.55	17.15	5.56	3.99	-1.45	18.60
B. As percentage of GDP								
1970	20.02	18.5	5.7	12.8	5.7	0.0	3.6	9.2
1971	22.00	17.9	5.1	12.8	5.1	0.0	3.1	9.8
1972	23.34	15.3	5.1	10.2	5.1	0.0	1.7	8.5
1973	25.64	16.9	4.7	12.1	4.7	0.0	3.3	8.8
1974	33.60	20.6	6.7	14.0	6.7	0.0	5.9	8.0
1975	36.39	25.2	12.2	12.9	12.2	0.0	0.4	12.6
1976	41.01	28.1	19.8	8.3	19.8	0.0	-1.6	9.9
1977	49.76	34.2	20.7	13.5	20.7	0.0	2.3	11.2
1978	55.15	25.4	12.0	13.4	12.0	0.0	0.5	12.9
1979	62.04	24.5	14.5	10.0	14.5	0.0	0.5	9.5
1980	70.16	22.6	11.5	11.1	8.3	3.2	1.5	9.6
1981	76.74	22.4	12.1	10.3	9.1	3.0	0.5	9.8
1982	90.09	23.3	13.4	9.9	8.4	5.0	-0.1	10.0
1983	94.63	20.9	10.5	10.4	6.0	4.5	-0.8	11.2
1984	104.81	21.8	8.5	13.3	3.6	4.9	0.3	13.0
1985	119.66	22.9	7.4	15.5	2.6	4.8	2.0	13.5
1986	134.33	20.3	4.1	16.2	1.8	2.3	0.7	15.4
1987	140.00	19.1	6.8	12.2	4.0	2.9	-1.0	13.3

Source: World Bank and author's estimates.

Table 6 Total Debt, Including Short-term Debt, 1970-87

Year	GDP	Debt	Debt/GDP
	In billions of dirham		
1970	20.02	n.a.	
1971	22.00	n.a.	
1972	23.34	3.95	16.92
1973	25.64	3.73	14.53
1974	33.60	4.29	12.76
1975	36.38	7.36	20.24
1976	41.01	10.59	25.83
1977	49.76	18.66	37.51
1978	55.15	21.98	39.85
1979	62.04	24.72	39.85
1980	70.16	42.03	59.90
1981	76.74	57.02	74.30
1982	90.09	77.75	86.30
1983	94.63	112.23	118.60
1984	104.83	133.13	127.00
1985	119.66	161.18	134.70
1986	134.33	164.02	122.10
1987	140.00	161.70	115.50

Source: IMF, World Bank, and author's estimates.

Table 7 Foreign Exchange Reserves of the Central Bank including Gold and SDRs

Year	Gold (US$)	Other reserves (US$)	Total reserves (US$)	DH/$	Total reserves Billions of dirham	Total imports Billions of dirham	Number of weeks of reserves
1970	21.00	119.00	140.00	5.03	0.70	4.32	8.47
1971	23.00	151.00	174.00	4.76	0.83	4.34	9.92
1972	23.00	204.00	227.00	4.67	1.06	4.48	12.29
1973	25.00	241.00	266.00	4.29	1.14	5.67	10.47
1974	26.00	391.00	417.00	4.15	1.73	9.45	9.53
1975	25.00	352.00	377.00	4.18	1.58	12.14	6.76
1976	24.00	467.00	491.00	4.48	2.20	15.55	7.36
1977	27.00	505.00	532.00	4.33	2.30	18.57	6.45
1978	31.00	618.00	649.00	3.89	2.52	16.77	7.82
1979	33.00	557.00	590.00	3.74	2.21	18.53	6.19
1980	29.00	399.00	428.00	4.33	1.86	20.66	4.67
1981	23.00	230.00	253.00	5.33	1.35	27.47	2.55
1982	20.00	218.00	238.00	6.27	1.49	31.31	2.48
1983	15.00	107.00	122.00	8.06	0.98	30.03	1.70
1984	13.00	49.00	62.00	9.55	0.59	38.67	0.80
1985	14.00	115.00	129.00	9.62	1.24	43.68	1.48
1986	14.00	211.00	225.00	8.71	1.96	42.72	2.39
1987	15.00	411.00	427.00	7.80	3.33	42.32	4.09

Source: IMF, *International Financial Statistics,* various years.

Note: Gold is valued at domestic prices and not at world prices. Total imports are of goods and nonfactor services.

Table 8 Foreign Assets and Liabilities of the Banking System, including to the IMF after 1980

Year	Central bank assets	Central bank liability	Central bank net	Deposit banks assets	Deposit banks liability	Deposit banks net	Total	Central bank	Deposit banks
				millions of dirham				billions of dirham	
1970	801	309	492	195	96	99	0.59	0.49	0.10
1971	919	88	831	199	69	130	0.96	0.83	0.13
1972	1,162	31	1,131	276	100	176	1.31	1.13	0.18
1973	1,241	26	1,215	315	77	238	1.45	1.22	0.24
1975	1,714	75	1,639	358	96	262	1.90	1.64	0.2
1976	2,276	735	1,541	394	120	274	1.82	1.54	0.2
1977	2,340	786	1,554	421	170	251	1.81	1.55	0.25
1978	2,584	1,098	1,486	492	160	332	1.82	1.49	0.33
1979	2,226	983	1,243	803	309	494	1.74	1.24	0.49
1980	1,871	2,399	-528	1,066	300	766	0.24	-0.53	0.77
1981	1,366	3,688	-2,322	1,737	360	1,377	-0.95	-2.32	1.38
1982	1,507	6,443	-4,936	1,769	217	1,552	-3.38	-4.94	1.55
1983	1,001	8,967	-7,966	2,148	303	1,845	-6.12	-7.97	1.85
1984	612	10,519	-9,907	2,432	323	2,109	-7.80	-9.91	2.11
1985	1,247	12,664	-11,417	3,100	472	2,628	-8.79	-11.42	2.63
1986	1,987	9,964	-7,977	3,378	740	2,638	-5.34	-7.98	2.64
1987	3,351	9,284	-5,933	n.a.	n.a.	2,879	-3.05	-5.93	2.88

Source: IMF, *International Financial Statistics*; Bank Al Maghrib; and World Bank.

Table 9 Index of Terms of Trade and EXchange Rates, 1967-82 (1980 = 100)

Year	Nominal exchange rate	Real exchange rate	Terms of trade ITPA
1967	91.76	85.21	120.07
1968	93.58	78.33	111.14
1969	95.11	76.98	110.25
1970	96.87	80.31	105.29
1971	95.93	80.89	114.14
1972	97.44	81.40	100.22
1973	99.00	87.89	100.88
1974	100.41	85.62	134.84
1975	99.10	89.30	140.46
1976	97.64	86.70	106.84
1977	98.02	92.06	100.88
1978	97.83	94.30	95.59
1979	97.54	92.92	105.51
1980	100.00	100.00	100.00
1981	93.93	96.33	102.65
1982	92.70	93.61	101.76

Source: World Bank, 1984a.

Table 10 Index of Terms of Trade and Exchange Rates, 1977-87 (1980 = 100)

Year	Nominal exchange rate	Real exchange rate
1977	101.00	106.50
1978	103.40	110.20
1979	102.80	106.60
1980	100.00	100.00
1981	93.70	98.00
1982	92.80	99.00
1983	88.30	96.20
1984	84.10	90.30
1985	78.00	86.30
1986	74.70	88.40
1987	71.50	86.60

Source: World Bank and author's estimates.

Table 11 Government Receipts and Expenditures Including Amortization of External Debt, With and Without Rescheduling of Interest

Year	Gross domestic product	Government receipts	Government expenditure, inc. amortization of principal — Before relief of principal and interest	Government expenditure, inc. amortization of principal — After relief of principal and interest	Government expenditure, exc. amortization of principal — Before relief of interest	Government expenditure, exc. amortization of principal — After relief of interest
A. In billion Dh						
1970	20.02		0.00	0.00	0.00	0.00
1971	22.00	3.27	4.40	4.40	4.23	4.23
1972	23.34	3.36	4.76	4.76	4.54	4.54
1973	25.64	4.14	5.07	5.07	4.83	4.83
1974	33.60	7.09	8.95	8.95	8.71	8.71
1975	36.39	9.49	12.02	12.02	11.80	11.80
1976	41.01	8.32	16.36	16.36	16.11	16.11
1977	49.76	10.78	19.99	19.99	19.56	19.56
1978	55.15	11.69	17.80	17.80	17.05	17.05
1979	62.04	13.08	22.10	22.10	21.08	21.08
1980	70.16	15.19	26.42	26.42	23.80	23.80
1981	76.74	17.84	32.07	32.07	28.76	28.76
1982	90.09	20.48	37.47	37.47	33.37	33.37
1983	94.63	21.08	37.28	32.69	33.48	32.57
1984	104.81	23.47	44.08	35.39	35.81	33.96
1985	119.66	26.75	48.93	41.72	38.76	37.65
1986	134.33	29.15	53.96	41.30	37.33	35.91
1987	140.00	32.75	54.47	44.58	41.79	39.65
B. As a percentage of GDP						
1970		0.00	0.00	0.00	0.00	0.00
1971		14.85	20.02	20.02	19.21	19.21
1972		14.40	30.39	20.39	19.45	19.45
1973		16.15	19.78	19.78	18.85	18.85
1974		21.11	26.63	26.63	25.92	25.92
1975		26.08	33.03	33.03	32.41	32.41
1976		20.29	39.89	39.89	39.29	39.29
1977		21.67	40.17	40.17	39.30	39.30
1978		21.20	32.27	32.27	30.92	30.92
1979		21.09	35.61	35.61	33.98	33.98
1980		21.65	37.65	37.65	33.93	33.93
1981		23.24	41.79	41.79	37.48	37.48
1982		22.73	41.59	41.59	37.04	37.04
1983		22.28	39.39	34.54	35.38	34.42
1984		22.39	42.06	33.77	34.17	32.04
1985		22.35	40.89	34.87	32.39	31.46
1986		21.70	40.17	30.75	27.79	26.73
1987		23.39	38.91	31.84	29.85	28.32

Source: World Bank.

Table 12 Total Government Expenditure by Category, 1971-1987, Noninterest, Capital and Debt Service With and Without Interest Rescheduling

Year	Gross domestic product	Current expenses after int.relief	Wages	Interest total	Current expenses except interest	Capital outlays	Debt before relief	Service after relief
A. In billion Dh								
1970	20.02						0.00	0.00
1971	22.00	3.11	1.84	0.22	2.88	1.12	0.40	0.40
1972	23.34	3.35	2.10	0.27	3.09	1.19	0.48	0.48
1973	25.64	3.62	2.21	0.28	3.35	1.21	0.52	0.52
1974	33.60	6.47	2.85	0.29	6.19	2.24	0.52	0.52
1975	36.39	7.35	3.28	0.34	7.00	4.45	0.57	0.57
1976	41.01	7.99	4.14	0.49	7.50	8.12	0.74	0.74
1977	49.76	9.25	5.11	0.75	8.50	10.31	1.18	1.18
1978	55.15	10.42	5.58	1.05	9.37	6.63	1.79	1.79
1979	62.04	12.07	6.65	1.36	10.71	9.01	2.37	2.37
1980	70.16	15.73	7.91	1.88	13.85	8.07	4.50	4.50
1981	76.74	19.44	9.33	3.07	16.37	9.32	6.38	6.38
1982	90.09	21.28	10.42	3.35	17.93	12.09	7.45	7.45
1983	94.63	22.61	11.42	3.91	18.71	9.96	8.61	4.94
1984	104.81	25.10	12.07	4.49	20.61	8.86	14.61	7.77
1985	19.66	28.79	12.65	6.95	21.84	8.86	18.23	12.14
1986	134.33	30.39	14.20	7.75	22.64	5.52	25.79	14.56
1987	140.00	30.10	15.39	7.72	22.38	9.55	22.54	14.79
B. As a percentage of GDP								
1970	20.02	0.00	0.00	0.00	0.00	0.00	0.00	0.00
1971	22.00	14.12	8.38	1.01	13.11	5.09	1.82	1.82
1972	23.34	14.35	9.00	1.14	13.22	5.10	2.07	2.07
1973	25.64	14.13	8.60	1.08	13.05	4.72	2.01	2.01
1974	33.60	19.26	8.49	0.85	18.41	6.67	1.55	1.55
1975	36.39	20.18	9.01	0.95	19.24	12.23	1.57	1.57
1976	41.01	19.49	10.10	1.20	18.28	19.80	1.81	1.81
1977	49.76	18.58	10.27	1.50	17.08	20.72	2.37	2.37
1978	55.15	18.89	10.12	1.90	17.00	12.02	3.25	3.25
1979	62.04	19.46	10.72	2.19	17.27	14.52	3.82	3.82
1980	70.16	22.43	11.28	2.68	19.74	11.50	6.41	6.41
1981	76.74	25.33	12.16	4.00	21.34	12.14	8.31	8.31
1982	90.09	23.62	11.57	3.72	19.90	13.42	8.27	8.27
1983	94.63	23.89	12.07	4.13	19.77	10.53	9.10	5.22
1984	104.81	23.95	11.52	4.28	19.67	8.45	13.94	7.42
1985	119.66	24.06	10.57	5.81	18.25	7.40	15.24	10.14
1986	134.33	22.62	10.57	5.77	16.86	4.11	19.20	10.84
1987	140.00	21.50	10.99	5.51	15.99	6.82	16.10	10.56

Source: World Bank.

Table 13 Total Government Recurrent Expenditures, Including Interest After Rescheduling, and Noninterest Expenditures by Category, 1971-87

Year	Gross domestic product	Total recurrent govt. expenditure of which	Interest payments	Noninterest expenditures, of which	Wages	Subsidies and transfers	Goods and services
A. In billion Dh							
1971	22.00	3.11	0.22	2.88	1.84	0.13	0.91
1972	23.34	3.35	0.27	3.09	2.10	0.14	0.84
1973	25.64	3.62	0.28	3.35	2.21	0.21	0.94
1974	33.60	6.47	0.29	6.19	2.85	2.09	1.25
1975	36.39	7.35	0.34	7.00	3.28	2.085	1.68
1976	41.01	7.99	0.49	7.50	4.14	1.38	1.98
1977	49.76	9.25	0.75	8.50	5.11	1.31	2.08
1978	55.15	10.42	1.05	9.37	5.58	1.10	2.69
1979	62.04	12.07	1.36	10.71	6.65	1.37	2.69
1980	70.16	15.73	1.88	13.85	7.91	2.29	3.65
1981	76.74	19.44	3.07	16.37	9.33	2.93	4.12
1982	90.09	21.28	3.35	17.93	10.42	2.87	4.65
1983	94.63	22.61	3.91	18.71	11.42	2.30	4.99
1984	104.81	25.10	4.49	20.61	12.07	2.82	5.72
1985	119.66	28.79	6.95	21.84	12.65	3.46	5.73
1986	134.33	30.39	7.75	22.64	14.20	2.56	5.89
1987	140.00	30.10	7.72	22.38	15.39	1.48	5.51
B. As percentage of GDP							
1970	20.02	0.00	0.00	0.00	0.00		
1971	22.00	14.12	1.01	13.11	8.38	0.60	4.14
1972	23.34	14.35	1.14	13.22	9.00	0.61	3.61
1973	25.64	14.13	1.08	13.05	8.60	0.80	3.65
1974	33.60	19.26	0.85	18.41	8.49	6.21	3.71
1975	36.39	20.18	0.95	19.24	9.01	5.62	4.61
1976	41.01	19.49	1.20	18.28	10.10	3.35	4.84
1977	49.76	18.58	1.50	17.08	10.27	2.63	4.17
1978	55.15	18.89	1.90	17.00	10.12	1.99	4.88
1979	62.04	19.46	2.19	17.27	10.72	2.21	4.33
1980	70.16	22.43	2.68	19.74	11.28	3.26	5.20
1981	76.74	25.33	4.00	21.34	12.16	3.82	5.36
1982	90.09	23.62	3.72	19.90	11.57	3.18	5.16
1983	94.63	23.89	4.13	19.77	12.07	2.43	5.27
1984	104.81	23.95	4.28	19.67	11.52	2.69	5.46
1985	119.66	24.06	5.81	18.25	10.57	2.89	4.79
1986	134.33	22.62	5.77	16.86	10.57	1.90	4.38
1987	140.00	21.50	5.51	15.99	10.99	1.06	3.93

Source: World Bank.

Table 14 External Factor Service Balance Including Worker Remittances, 1970-87

Year	GDP	Receipts including remittances	Expenses including interest
A. In billion Dh			
1970	20.02	0.39	0.47
1971	22.00	0.55	0.48
1972	23.34	0.70	0.53
1973	25.64	1.08	0.51
1974	33.60	1.67	0.51
1975	36.39	2.27	0.56
1976	41.01	2.52	0.76
1977	49.76	2.79	1.09
1978	55.15	3.29	1.55
1979	62.04	3.85	2.01
1980	70.16	4.30	2.62
1981	46.74	5.43	4.06
1982	90.09	5.27	4.29
1983	94.63	6.59	4.73
1984	104.81	7.83	5.45
1985	119.66	9.89	8.04
1986	134.33	12.75	6.55
1987	140.00	13.40	6.68
B. As percentage of GDP			
1970	20.02	1.95	2.33
1971	22.00	2.50	2.20
1972	23.34	3.00	2.25
1973	25.64	4.21	2.00
1974	33.60	4.98	1.52
1975	36.39	6.24	1.54
1976	41.01	6.15	1.84
1977	49.76	5.61	2.18
1978	55.15	5.96	2.82
1979	62.04	6.20	3.25
1980	70.16	6.12	3.73
1981	76.74	7.07	5.29
1982	90.09	5.85	4.76
1983	94.63	6.96	5.00
1984	104.81	7.47	5.20
1985	119.66	8.26	6.72
1986	134.33	9.49	4.87
1987	140.00	9.57	4.77

Source: World Bank and author's estimates.

Table 15 Exports of Goods and Nonfactor Services, 1970-87

Year	GDP	Goods	Nonfactor services	Total
A. In billion Dh				
1970	20.02	2.46	1.07	3.53
1971	22.00	2.52	1.12	3.64
1972	23.34	2.95	1.39	4.34
1973	25.64	3.75	1.60	5.35
1974	33.60	7.44	18.85	9.29
1975	36.39	6.18	1.99	8.17
1976	41.01	5.58	2.08	7.66
1977	49.76	5.86	2.63	8.49
1978	55.15	6.26	2.83	9.09
1979	62.04	7.71	2.91	10.62
1980	70.16	9.68	3.34	13.03
1981	76.74	12.02	4.08	16.10
1982	90.09	12.51	5.51	18.02
1983	94.63	15.08	6.30	21.38
1984	104.81	19.41	7.42	26.83
1985	119.66	22.12	10.01	32.13
1986	134.33	22.61	10.46	33.07
1987	140.00	24.04	11.49	35.53
B. As percentage of GDP				
1970	20.02	12.29	5.33	17.62
1971	22.00	11.46	5.10	16.55
1972	23.34	12.63	5.95	18.58
1973	25.64	14.61	6.25	20.85
1974	33.60	22.15	5.49	27.64
1975	36.39	16.99	5.46	22.45
1976	41.01	13.61	5.06	18.67
1977	49.76	11.77	5.28	17.05
1978	55.15	11.35	5.13	16.48
1979	62.04	12.43	4.69	17.12
1980	70.16	13.80	4.76	18.56
1981	76.74	15.66	5.32	20.98
1982	90.09	13.88	6.12	20.00
1983	94.63	15.94	6.66	22.60
1984	104.81	18.52	7.08	25.59
1985	119.66	18.49	8.37	26.85
1986	134.33	16.83	7.79	24.62
1987	140.00	17.17	8.21	25.38

Source: World Bank and author's estimate.

Table 16 Exports of Good by Major Category Including Subcontracting, 1970-87

Year	GDP	Energy	Phosphates derivatives minerals	Agriculture food tobacco	Nonfood manufacture including subcontracting	Total goods
A. In billion Dh						
1970	20.02	0.00	0.82	1.41	0.23	2.46
1971	22.00	0.00	0.84	1.36	0.32	2.52
1972	23.34	0.00	0.95	1.64	0.36	2.95
1973	25.64	0.02	1.13	2.12	0.48	3.75
1974	33.60	0.05	4.65	2.02	0.71	7.44
1975	36.39	0.06	3.79	1.64	0.69	6.18
1976	41.01	0.08	2.62	2.08	0.80	5.58
1977	49.46	0.09	2.85	1.93	0.98	5.86
1978	55.15	0.09	2.85	2.23	1.10	6.26
1979	62.04	0.28	3.50	2.50	1.43	7.71
1980	70.16	0.47	4.66	3.00	1.56	9.68
1981	76.74	0.54	6.12	3.34	2.02	12.02
1982	90.09	0.53	6.17	3.29	2.52	12.51
1983	94.63	0.58	7.26	4.10	3.14	15.08
1984	104.81	0.76	9.79	4.74	4.12	19.41
1985	119.66	0.81	10.10	58.81	5.41	22.12
1986	134.33	0.56	8.37	7.21	6.47	22.61
1987	140.00	0.64	8.25	7.17	7.97	24.04
B. As percentage of GDP						
1970	20.02	0.00	4.09	7.06	1.14	12.29
1971	22.00	0.00	3.80	6.19	1.47	11.46
1972	23.34	0.00	4.05	7.04	1.54	12.63
1973	25.64	0.08	4.39	8.28	1.86	14.61
1974	33.60	0.15	13.85	6.02	2.13	22.15
1975	36.39	0.16	10.42	4.52	1.90	16.99
1976	41.01	0.19	6.40	5.07	1.95	13.61
1977	49.76	0.18	5.73	3.88	1.98	11.77
1978	55.15	0.16	5.17	4.03	1.99	11.35
1979	52.04	0.44	5.64	4.03	2.31	12.43
1980	70.16	0.67	6.64	4.27	2.23	13.80
1981	76.74	0.71	7.97	4.35	2.63	15.66
1982	90.09	0.59	6.85	3.65	2.80	13.88
1983	94.63	0.62	7.67	4.33	3.32	15.94
1984	104.81	0.72	9.34	4.52	3.93	18.52
1985	119.66	0.68	8.44	4.85	4.52	18.49
1986	134.33	0.42	6.23	5.37	4.82	16.83
1987	140.00	0.46	5.90	5.12	5.70	17.17

Source: World Bank and author's estimate.

Table 17 Imports of Goods and Nonfactor Services, 1970-87

Year	GDP	Goods nongovernment	Goods government	Total goods and nonfactor services
A. In billion Dh				
1970	20.02	3.47	0.35	4.32
1971	22.00	3.60	0.31	4.34
1972	23.34	3.73	0.31	4.48
1973	25.64	4.77	0.35	5.67
1974	33.60	8.34	0.45	9.45
1975	36.39	10.36	1.06	12.14
1976	41.01	11.80	3.21	15.55
1977	49.76	14.58	3.23	18.57
1978	55.15	12.57	3.24	16.77
1979	62.04	14.43	3.22	18.53
1980	70.16	16.79	2.82	20.66
1981	76.74	22.45	3.61	27.47
1982	90.09	25.99	3.76	31.31
1983	94.63	25.59	2.84	30.03
1984	104.81	34.40	2.67	38.67
1985	118,66	38.61	3.09	43.68
1986	134.33	34.61	6.05	42.72
1987	140.00	35.27	4.59	42.32
B. As percentage of GDP				
1970	20.02	17.34	1.76	21.58
1971	22.00	16.38	1.42	19.73
1972	23.34	15.97	1.33	19.19
1973	25.64	18.60	1.35	22.11
1974	33.60	24.81	1.34	28.13
1975	36.39	28.48	2.91	33.36
1976	41.01	28.77	7.82	37.92
1977	49.76	29.29	6.49	37.32
1978	55.15	22.80	59.87	30.41
1979	62.04	23.26	5.20	29.87
1980	70.16	23.93	4.02	29.45
1981	76.74	29.26	4.71	35.80
1982	90.09	28.85	4.17	34.75
1983	94.63	27.04	3.00	31.73
1984	104.81	32.82	2.55	36.90
1985	119.66	32.27	2.58	36.50
1986	134.33	25.76	4.50	31.80
1987	140.00	25.19	3.28	30.23

Source: World Bank and author's estimate.

Table 18 Imports of Goods by Principal Category, 1970-87

Year	GDP	Food	Fuel	Total intermediate goods	Capital	Consumer goods	Total
A. In billion Dh							
1970	20.02	0.58	0.19	1.30	0.83	0.57	3.47
1971	22.00	0.70	0.23	1.33	0.79	0.55	3.60
1972	23.34	0.62	0.26	1.56	0.71	0.57	3.73
1973	25.64	1.03	0.30	1.93	0.85	0.66	4.77
1974	33.60	1.83	1.13	3.15	1.42	0.80	8.34
1975	36.39	2.59	1.12	2.97	2.49	1.19	10.36
1976	41.01	1.98	1.30	3.65	3.44	1.43	11.80
1977	49.76	1.94	1.67	4.53	4.97	1.46	14.58
1978	55.15	2.01	1.78	4.13	3.39	1.27	12.57
1979	62.04	2.14	2.77	5.05	3.29	1.18	14.43
1980	70.16	2.83	3.96	5.51	3.17	1.32	16.79
1981	76.74	4.61	6.12	6.35	3.86	1.51	22.45
1982	90.09	3.50	7.07	7.88	5.74	1.82	25.99
1983	94.63	3.80	7.03	8.14	4.85	1.77	25.59
1984	104.81	5.82	8.99	11.04	6.46	2.09	34.40
1985	119.66	5.04	10.81	13.47	6.53	2.76	38.61
1986	134.33	4.33	5.43	13.13	8.25	3.47	34.61
1987	140.00	3.98	6.17	13.65	7.36	4.11	35.27
B. As percentage of GDP							
1970	20.02	2.92	0.94	6.49	4.16	2.83	17.34
1971	22.00	3.19	1.06	6.04	3.59	2.50	16.38
1972	23.34	2.67	1.10	6.70	3.04	2.46	15.97
1973	25.64	4.03	1.18	7.51	3.32	2.56	18.60
1974	33.60	5.46	3.35	9.38	4.24	2.38	24.81
1975	36.39	7.13	3.08	8.17	6.85	3.26	28.48
1976	41.01	4.82	3.18	8.89	8.40	3.49	28.77
1977	49.76	3.91	3.36	9.10	9.99	2.94	29.29
1978	55.15	3.64	3.23	7.49	6.14	2.30	22.80
1979	62.04	3.45	4.46	8.14	5.30	1.90	23.26
1980	70.16	4.04	5.65	7.86	4.52	1.87	23.93
1981	76.74	6.01	7.98	8.27	5.03	1.97	29.26
1982	90.09	3.88	7.84	8.74	6.37	2.01	28.85
1983	94.63	4.01	7.43	8.60	5.13	1.87	27.04
1984	104.81	5.55	8.58	10.53	6.16	2.00	32.82
1985	119.66	4.22	9.03	11.26	5.46	2.30	32.27
1986	134.33	3.22	4.04	9.77	6.14	2.58	25.76
1987	140.00	2.84	4.41	9.75	5.26	2.94	25.19

Source: World Bank and author's estimate.

Table 19 Structure of Tax Revenues, 1972-87

Year	Gross domestic product	Direct taxes	Property transfers	Domestic indirect taxes	Foreign trade	Other	Nontax revenues	Petroleum levy
A. In billion Dh								
1972	23.35	0.66	0.10	1.21	1.08	0.13	9.19	
1973	25.64	0.82	0.12	1.31	1.34	0.16	0.40	
1974	33.60	1.11	0.13	1.38	2.12	0.19	2.16	
1975	36.39	2.56	0.15	1.62	2.62	0.16	1.39	
1976	41.01	1.65	0.21	1.92	2.95	0.19	1.41	
1977	49.76	2.56	0.29	2.56	4.17	0.25	0.96	
1978	55.15	29.95	0.26	3.00	4.17	0.28	1.03	
1979	62.04	3.33	0.27	3.33	5.03	0.31	1.54	
1980	70.16	3.34	0.28	4.07	5.86	0.34	1.31	
1981	76.74	3.77	0.27	4.00	6.93	0.35	2.52	
1982	90.09	3.83	0.30	4.99	8.61	0.40	2.34	
1983	94.64	4.33	0.35	5.82	9.14	0.46	2.00	
1984	104.80	4.99	0.39	5.94	9.25	0.61	2.30	
1985	119.66	5.48	0.47	6.75	9.86	0.67	3.37	
1986	134.33	6.23	0.52	6.78	9.76	0.77	1.32	3.77
1987	140.00	7.09	0.63	7.73	10.04	0.81	2.05	4.41
B. As percentage of GDP								
1972	23.35	2.81	0.41	5.18	4.63	0.54	0.83	0.00
1973	25.64	3.21	0.45	5.11	5.22	0.62	1.55	0.00
1974	33.60	3.29	0.39	4.10	6.32	0.58	6.43	0.00
1975	36.39	7.02	0.40	4.44	7.21	0.45	3.81	0.00
1976	41.01	4.01	0.50	4.67	7.20	0.46	3.45	0.00
1977	49.76	5.14	0.58	5.14	8.38	0.51	1.93	0.00
1978	55.15	5.34	0.48	5.44	7.57	0.50	1.87	0.00
1979	62.04	5.36	0.44	5.36	8.10	0.50	2.48	0.00
1980	70.16	4.76	0.40	5.80	8.35	0.49	1.86	0.00
1981	76.74	4.91	0.35	5.22	9.03	0.46	3.28	0.00
1982	90.09	4.25	0.34	5.54	9.56	0.45	2.60	0.00
1983	94.64	5.58	0.37	6.15	8.60	0.48	2.11	0.00
1984	104.80	4.76	0.37	5.66	8.83	0.58	2.19	0.00
1985	119.66	4.58	0.39	5.64	8.24	0.56	2.81	0.00
1986	134.33	4.64	0.39	5.04	7.27	0.57	0.98	2.81
1987	140.00	5.06	0.45	5.52	7.17	0.58	1.46	3.15

Source: World Bank and author's estimates.

Table 20 Indirect Taxes and Consumption Subsidies, 1972-87

Year	Gross domestic product	Indirect taxes (including on imports)	Consumption subsidies payment order basis	Taxes, net of consumption subsidies
A. In billion Dh				
1972	23.35	1.73	0.00	1.73
1973	25.64	1.91	0.14	1.78
1974	33.60	2.28	1.58	0.70
1975	36.39	2.66	1.66	1.01
1976	41.01	3.15	0.94	2.21
1977	49.76	4.14	0.70	3.44
1978	55.15	4.42	0.39	4.03
1979	62.04	4.96	0.61	4.35
1980	70.16	5.81	1.68	4.13
1981	76.74	6.04	2.37	3.67
1982	90.09	7.81	2.32	5.49
1983	94.64	8.69	2.03	6.66
1984	104.80	9.56	2.03	7.53
1985	119.66	10.78	3.35	7.43
1986	134.33	14.71	1.53	13.18
1987	140.00	16.53	0.63	15.90
B. As percentage of GDP				
1972	23.35	7.40	0.00	7.40
1973	25.64	7.45	0.53	6.93
1974	33.60	6.80	4.70	2.10
1975	36.39	7.32	4.55	2.77
1976	41.01	7.69	2.29	5.40
1977	49.76	8.33	1.40	6.92
1978	55.15	8.01	0.71	7.31
1979	62.04	7.99	0.98	7.00
1980	70.16	8.28	2.39	5.89
1981	76.74	7.87	3.09	4.78
1982	90.09	8.67	2.58	6.09
1983	94.64	9.18	2.15	7.04
1984	104.80	9.12	1.94	7.18
1985	119.66	9.01	2.80	6.21
1986	134.33	10.95	1.14	9.81
1987	140.00	11.81	0.45	11.36

Source: World Bank and author's estimates.